Mine to Market

This book examines the practical and economic logic behind mining project evaluation, bridging the gap between technical quality and finance ability. It provides a comprehensive framework that connects geology, mining, processing, product marketability, logistics, costs, cash flow, risk, and project maturity. By focusing on real-world decision-making, it equips professionals to understand how a mineral resource becomes, or fails to become, a financeable mining project. It also offers practical tools to improve project screening, technical review, and due diligence processes.

- Explains how geology, processing, logistics, and finance ability interact to shape project outcomes.
- Provides tools like checklists, due diligence questions, and rapid project-reading logic for immediate application.
- Highlights the distinction between geological attractiveness and real economic strength.
- Demonstrates how project maturity, sensitivity, and sequencing affect investment credibility.
- Includes applied calculation methods to test assumptions and challenge unrealistic presentations.

The book is intended for mining professionals, project evaluators, and decision-makers seeking a clear, practical framework for assessing mining projects.

Mine to Market

Practical Frameworks for Mining Project Evaluation

Georg Wall

CRC Press
Taylor & Francis Group
Boca Raton London New York

CRC Press is an imprint of the
Taylor & Francis Group, an **informa** business

Designed cover image: Georg Wall

First edition published 2027
by CRC Press
2385 NW Executive Center Drive, Suite 320, Boca Raton FL 33431

and by CRC Press
4 Park Square, Milton Park, Abingdon, Oxon, OX14 4RN

CRC Press is an imprint of Taylor & Francis Group, LLC

ISBN: 9781041397625 (hbk)
ISBN: 9781041396291 (pbk)
ISBN: 9781042018932 (ebk)

DOI: 10.1201/9781042018932

Typeset in Times
by KnowledgeWorks Global Ltd.

Contents

Author Biography

Georg Wall, CEng, is a mineral processing and mining engineer with extensive expertise in mineral project evaluation, technical due diligence, techno-economic analysis, operational improvement, and project governance. His interdisciplinary background combines university studies in mineral processing and mining engineering with practical technical experience in software development. He is a Professional Member of the Institute of Materials, Minerals and Mining (MIMMM, UK), and Qualified for Minerals Reporting (QMR). Over more than two decades, he has contributed to international mining assignments across different commodities, jurisdictions, project stages, and operating environments. His work has included geological and mine modelling, CAPEX/OPEX estimation, strategic mine planning, technical due diligence, operational improvement, and technical governance. He has held senior roles as independent consultant, lead engineer, chief mining engineer, project manager, and vice president of operations, working across Europe, Africa, Asia, South America, and the CIS region. His work focuses on the integrated logic of mining projects, linking geology, mining, processing, product quality, logistics, economics, risk, and investment logic to deliver practical solutions in mining projects.

Preface

Mining rarely forgives superficial understanding. A project may have strong geology and still fail to become a strong business. It may show attractive grades and still produce a weak product. It may look convincing in a model and yet later reveal that the essential coherence between resource, processing, market, logistics, and capital was never truly there. Without that coherence, a mineral deposit does not become a business. It remains only a possibility.

That tension is where this book comes from.

Mine to Market was not written as another narrowly technical textbook on a single discipline, nor as an attempt to replace the specialist literature on geology, mine planning, mineral processing, or finance. Its purpose is different. It aims to bring together, within one practical logic, elements that in real work are too often separated. A geologist may see mineralization. A mining engineer may see extractability. A metallurgist may see recovery. An economist may see cash flow. An investor may see risk and return. Yet, value in mining is not created inside any one of these disciplines alone. It appears only when they begin to function as one coherent system.

For that reason, this book begins not with the market and not with money, but at the earliest point: the moment when mineralization becomes an object of evaluation at all. From there, it follows the chain step by step: from observation and interpretation to the geological model, from the model to the mineral resource, from the resource to extractability, product quality, revenue, costs, time, mining sequence, project maturity, and ultimately to the project's ability to stand before capital. In that sense, the book treats a mining project not as a set of reports and tables, but as a system for converting a natural occurrence into an economic result.

But in real work, general logic is never enough on its own.

At some point one has to stop trusting elegant language, open the spreadsheet, take out the calculator, and test what actually stands behind the assumptions. That is why the practical part of this book matters so much. The checklists, due diligence questions, discussion of typical errors, simple calculations, and applied cases are not included for volume, and not for methodological completeness as such. Their purpose is more practical than that: to help the reader see project structure faster, identify weak links earlier, ask better questions, and distinguish between a project that merely looks attractive and one that is genuinely robust.

In truth, that practical part has no natural end. Any experienced practitioner knows that the list of useful questions can be extended almost without limit. One can always add another checklist, another scenario, another sensitivity test, another example of how a project changes when price, recovery, logistics, product quality, or mining sequence change. In that sense, mining practice is inherently open-ended: it is always wider than any one book. That is precisely why the aim here was not to cover everything, but to offer a working framework that the reader can continue using beyond the last page.

I have deliberately not tried to make this book "final." In mining, that would be the wrong promise. Too much depends on deposit type, project stage, jurisdiction, market structure, process route, infrastructure, and the quality of the underlying data. What I wanted instead was something more useful: a book that helps people think correctly. A book that does not replace professional judgment, but disciplines it. A book that shows that mining errors do not arise only in formulas, but in the transitions between disciplines: between geology and resource estimation, between resource and product, between product and market, between revenue and cash flow, and between technical attractiveness and financeability.

This book is written for those who approach mining projects from different sides: geologists and mining engineers, metallurgists and planners, analysts, managers, investors, advisers, and also professionals in management companies and corporate functions who need to understand not only their own part of the work, but the logic of the system as a whole. One of the industry's persistent problems is that different participants in the same project often speak different professional languages, while the project itself remains one integrated reality.

If, after reading this book, the reader sees more clearly where value is actually created, where it is lost, which assumptions truly matter, and which merely create an appearance of strength, then the book will have done its job.

A mining project begins in the ground. But its fate is not decided there. It is decided in the quality of the data, the honesty of interpretation, the realism of the process route, the discipline of costs, the understanding of the market, the sequence of mining, the ability to withstand doubt, and the discipline to turn complexity into grounded decision-making.

That is what this book is about.

Declaration on language support

The author is not a native English speaker. Hence, in preparing the English manuscript, digital language tools, including ChatGPT, DeepL Translator, and Google Translate, were used as support for translation, comparison of wording, language editing and consistency of terminology. The content, structure, technical arguments, examples, calculations, professional judgement, and final text remain the author's own.

Acknowledgements

This book has been shaped not only by my own professional experience, but also by the individuals who taught me how to think across disciplines. I am deeply grateful to my teachers, professors, mentors, and colleagues, whose knowledge, expertise, and professional generosity have extremely influenced my perspective on mining projects.

In particular, I would like to acknowledge B. N. Kravets, I. G. Tarchevskaya, E. E. Ivanov, M. Scherschel, U. Lenski, R. Schaeffer, W. Thiels, A. Emons, L. Hoffmann, and R. Braun.

Some of them provided me with direct advice/assistance in geology, mineral processing, mining engineering, economics, and management. Others influenced me through discussions, project reviews, practical work, and leading by example. Without them, the integrated thinking framework presented in this book, especially the analysis of interdependencies between geology, mining, processing, product, market, economics, and capital, would not have evolved into its current form.

I also wish to thank my family. Larissa provided the patience and stability necessary to see this work through to completion. Alissa and Arthur inspired me in a different way: by reminding me that knowledge only matters when it can be explained clearly and with purpose.

1 Foundational Concepts *From Mineralization to Resource*

1.1 FROM GEOLOGICAL INDICATION TO EVALUATION OBJECT

1.1.1 When Mineralization Becomes an Object of Evaluation

When people speak about mineral deposits, they often create the impression that these exist in nature as fully defined entities, with clear boundaries, known tonnage, and established grade. In reality, nothing of that kind exists in the subsurface. What exists are geological processes that may, under favourable conditions, concentrate minerals. Everything beyond that is the result of observation, interpretation, and the progressive reduction of uncertainty.

This is the first point to understand: a deposit does not appear in ready-made form. It takes shape gradually as data are accumulated and the geological model is refined.

At the early stages of exploration, all that is usually known is a set of isolated indications of mineralization. These may include a mineralized outcrop, a geochemical anomaly, a geophysical response, or a mineralized interval intersected by a drill hole, trench, test pit, etc.. Such observations are best described as mineral occurrences or indications of mineralization. They show that mineralizing processes have operated within a given geological system, but on their own they say little about scale, continuity, or economic relevance.

The mere presence of mineralization does not mean that there is a deposit, still minus a mineral asset.

Mineralization may prove to be nothing more than a local lens, an isolated vein, or a small discontinuous zone with no meaningful extension. It may be geologically interesting, yet too limited in size, too erratic in distribution, or too inconsistent in grade to become an object of sustained technical or commercial interest. The history of exploration is full of examples where impressive assays from individual samples, or a visually striking outcrop, ultimately turned out to represent no more than a minor occurrence.

That is why geology and the mining industry distinguish several levels of understanding. The starting point is a mineral occurrence, that is, evidence that mineralization is present. From there, attention may shift to a geologically interesting zone, where mineralization appears to have meaningful extent.

The next step is an exploration target, meaning a working hypothesis about the possible size and grade range of the mineralized material.

Only after a sufficient body of evidence has been assembled does the object become something that can be evaluated as a mineral resource, namely a quantified

DOI: 10.1201/9781042018932-1

concentration of mineralized material with reasonable prospects for eventual economic extraction.

This progression is what transforms a geological indication into an object of evaluation and, potentially, into a mineral asset.

At this stage, the task is not simply to establish that mineralization exists. The real task is to understand its geological character, its controls, and its likely continuity.

A sample, in essence, provides only a single point of information in space. It shows whether mineralization is present at a particular location and depth. But a sample by itself says very little about the scale or geometry of the mineralized system.

The key question, therefore, is how individual observations and samples relate to one another, and whether they can be interpreted as parts of a single mineralized body.

In some cases, this requires substantial drilling and sampling. In others, a relatively small number of data points may be sufficient, provided they are consistent with the geological model and with what is known about that deposit type.

Consider a simple example. The first drill hole intersects a thick mineralized interval. A second hole, drilled 1 or 2 kilometres along the interpreted strike of the structure, encounters essentially the same style of mineralization and a comparable interval thickness. Additional geophysical data define a continuous anomaly between the two intervals.

In that situation, it may be reasonable to infer that the mineralization continues between the holes over a significant distance. The basis for that inference is not the sheer number of holes or samples, but the degree to which the observations are coherent with the geological model for that type of mineral system.

In some settings, such an interpretation may be reasonably robust even on a relatively wide data spacing. Stratiform iron formations and coal basins, for example, often display a high degree of geological continuity. In such cases, the thickness and position of mineralized horizons may persist over long distances, allowing their geometry to be interpolated over kilometers from a limited number of control points.

In other deposit types, the situation is very different. Vein systems, skarns, and some epithermal environments may be highly heterogeneous. Mineralization can change sharply over short distances, which means that a much denser pattern of drilling and sampling is required before the geometry can be understood with confidence.

For that reason, this stage can never be reduced to a simple rule based on the number of drill holes or samples. The real question is always whether the available data are sufficient to support a geological interpretation that can be defended.

Once such an interpretation becomes possible, the spatial distribution of mineralization begins to take shape, including its extent, thickness, internal variability, and geological controls.

Even then, however, it is still premature to speak of a mineral asset in any meaningful technical or commercial sense.

For an object to be considered a resource, it is not enough merely to demonstrate the presence of mineralization. Its key parameters must also be quantified. One must understand the genesis of the mineralization, the way grade is distributed, the degree to which values can be interpolated between data points, and whether there are reasonable grounds to assume that the material may ultimately be extracted.

In international reporting practice, this stage of knowledge is described as a mineral resource, meaning mineralized material for which tonnage, grade, and geological continuity have been established, and for which there are reasonable prospects for eventual economic extraction. It is at that point that a geological object begins to acquire technical and economic significance.

This is an important distinction. The mining industry does not exist for minerals as such. The Earth's crust contains enormous volumes of mineralized rock, and most of that material will never become the basis of a mining operation.

Value arises only when mineralized material can be delineated with sufficient confidence, extracted, processed, and sold into a market.

1.1.2 Evaluation Logic and Sources of Geological Information

If mineralization in itself does not yet mean that a deposit exists, the next question is this: how does a mineralized occurrence actually become an object of serious evaluation?

Exploration does not begin with drilling. It begins with the assumption that a particular geological setting may have been favourable for the formation of mineralization.

That assumption rarely arises in a vacuum. It is usually grounded in regional geology, known deposit types, results from earlier work, geophysical or geochemical data, or some combination of these. In some cases, the starting point is an old report or archival material. In others, it is an observation made during geological traverses. Elsewhere, the first signal may come from surface samples collected during mapping, from trenches, pits, or isolated outcrops. Sometimes attention is drawn by a mineralized exposure, a strong anomaly, or the coincidence of several independent indicators.

At this stage, no one is yet speaking about a deposit, still less about a mineral asset. What exists is simply an area that appears to warrant testing. That is how an exploration target comes into being.

An exploration or evaluation target is not a resource and not confirmed mineralization. It is an area within which there are grounds to assume the possible presence of mineralization of a certain type, scale, or geometry. In other words, it is a working hypothesis translated into spatial limits and prepared for testing.

From that point onwards, exploration ceases to be a loose collection of observations and becomes a sequence of deliberate actions. Its purpose is not simply to gather more information, but to test how well the original hypothesis corresponds to the actual structure of the subsurface object.

That is the essence of exploration/evaluation strategy.

A sound exploration and evaluation strategy is built around a simple principle: at each stage, one should obtain the maximum amount of information that reduces uncertainty, while avoiding expenditure on work that does not materially improve understanding of the object. Exploration is almost always carried out under budget constraints, time pressure, and incomplete data. The real question, therefore, is not how to do more work, but how to select the work that adds the greatest value to the next decision.

In broad terms, the sequence is straightforward. First, the geological setting of the area is refined. Then the indicators that may point to mineralization are identified. After that, appropriate methods of testing are selected: traverses, mapping, sampling, geophysics, geochemistry, trenches, pits, and drilling. But that sequence alone does not guarantee success. What matters is not the list of methods, but the logic behind their use.

At the early stages, the main objective is to determine as cheaply as possible whether there is any basis for moving forward at all. For that purpose, comparatively fast and lower-cost methods are used. These may include follow-up mapping, surface sampling, reinterpretation of historic data, structural analysis, ground geophysics, or local geochemical work. At this level, the key issue is not detail, but the basic question: does the original idea hold, or does it not?

If the initial results are encouraging, the exploration programme becomes more focused. The task is no longer simply to record the presence of mineralization, but to understand its position, character, orientation, extent, and possible internal variability. At that stage, the role of trenches, pits, more detailed geophysics, and ultimately drilling becomes much more important.

Drilling occupies a special place in exploration strategy. On the one hand, it is one of the most expensive methods available. On the other, it provides the most valuable direct information about mineralization at depth. For that reason, drilling should never be treated as an end in itself. It should be used as a tool for testing a specific geological hypothesis.

Every drill hole should answer a defined question: Is the interpreted position of mineralization confirmed? Is thickness maintained? Does the style of mineralization match what was expected? Does grade vary along strike or down dip? If a hole is drilled simply in the hope of "hitting something," without a clear question behind it, that is no longer strategy. It is expenditure without discipline.

This is often where the line is drawn between systematic exploration and chaotic search. In systematic exploration, each subsequent activity is justified because the previous one narrowed uncertainty and made it possible to formulate a new question. In chaotic search, work is often carried out simply because "more checking is needed" or, in the worst case, because the budget must be spent.

This matters because most exploration areas do not become deposits, and even fewer become mineral assets of real technical and commercial significance. The reason is usually not the complete absence of mineralization. Mineralization may well be present, sometimes in a fairly convincing form. But it may prove too local, too discontinuous, too low in grade, or too structurally complex. Another possibility is that mineralization exists, but subsequent work shows that its scale or characteristics do not support any realistic expectation of future economic extraction.

For that reason, exploration always includes not only stages of confirmation, but also stages of rejection. That is a normal part of the process. A target that fails under testing should be stopped in time. That too is part of professional discipline. A good exploration strategy is not only about identifying promising objects, but also about screening out unpromising ones early enough.

In practice, this means that every stage should end with a decision point. One must either move forward, refine the model, or stop the work. Such decisions should not be based on intuition, nor on a single encouraging sample, but on the extent to which new data have materially changed the understanding of the object.

For example, if surface sampling confirms the presence of mineralization but does not demonstrate consistency over an area, additional mapping or trenching may be required. If drilling confirms the expected position of a mineralized zone and aligns with geophysical evidence, that may justify an expanded programme. If, however, new data contradict the original model, the correct response is not to keep drilling out of momentum, but to revisit the hypothesis.

That is why exploration strategy is not a list of methods and not a calendar of field activities. It is a system for the sequential testing of hypotheses.

First, one must understand why a particular area deserves attention at all. Then, which indicators genuinely point to mineralization. Then, whether that mineralization is consistent with the expected geological hypothesis or deposit model. Only after that does the question arise of its spatial development, continuity, and possible scale.

Depending on deposit type, this logic may take different forms. In stratiform and tabular systems, greater weight is placed on horizon continuity, stratigraphy, and areal persistence. In vein and skarn systems, structure, morphological variability, and local controls are critical. In large disseminated systems such as porphyries, it is often more important to identify the broader mineralized system, its zonation, and its scale than to focus too early on isolated high-grade intervals. But in all cases, the governing principle remains the same: each stage must deliver new information that moves the work from a general idea toward a defensible understanding of the object.

Good exploration, therefore, is not a matter of maximizing the number of drill holes, samples, or line-kilometres. Good exploration is the reduction of uncertainty at reasonable cost and within a reasonable timeframe.

That is also why the emergence of a serious mineral asset candidate is almost never the result of a single successful drill hole or one visually compelling anomaly. What matters is the point at which scattered observations begin to form a coherent picture, when it becomes clear that the mineralization is not accidental, not purely local, and not limited to isolated signs, but part of a system that can be studied systematically, interpreted, and ultimately evaluated on a quantitative basis.

From that point, the work moves into the next phase. The question is no longer simply whether mineralization exists, but whether the underlying data are of sufficient quality to support everything that follows.

And that leads directly to the next issue, one that matters in mining far more than is often assumed: sample quality, data reliability, data density, and above all, interpretation.

1.2 FROM DATA TO GEOLOGICAL INTERPRETATION

1.2.1 Interpretation of Geological Data

As exploration advances, information begins to accumulate. Sampling results become available, drilling intervals are recorded, geophysical data are added, and further observations are made through mapping and fieldwork.

Yet these data do not, by themselves, create an understanding of the object.

Each individual result reflects only a small part of the overall picture. A sample shows the grade of the valuable component at a specific point. A drill hole records the sequence of rock units and mineralization along its trace. Geophysical methods register the physical properties of rocks. Field observations help establish structure, the spatial relationships between rock units, and the possible controls on mineralization.

None of these sources of information, taken on its own, is sufficient to describe the object.

For that reason, the main task at this stage is not simply to accumulate more data but also to correlate them.

Individual observations must be considered within a single interpretive framework. Mineralized intervals identified in drilling are compared with the geological structure of the area. Surface samples are checked against drilling results. Geophysical anomalies are tested against geological observations and sampling data.

It is through this process of correlation that an understanding of how mineralization is distributed gradually begins to emerge.

At the same time, it is important to remember that every sample or observation is only a single point of information. Even a high grade in an individual sample says nothing on its own about the scale of mineralization. It may prove to be only a local occurrence with no continuation.

For that reason, interpretation is never built on isolated results. It develops only when several independent observations begin to form a coherent picture.

In some cases, the data support one another. Mineralized intervals in drilling coincide with geophysical anomalies, and their position is consistent with the geological structure. In such cases, confidence in the interpretation gradually increases.

In other situations, the data may conflict. Geophysics may define an extensive anomaly, while drilling fails to confirm mineralization. Or surface samples may return high grades that are not repeated in drill holes. Such discrepancies are common enough and usually require the original assumptions to be revisited.

It is at this stage that geological interpretation begins to play a central role.

The interpretation of geological information is one of the most critical stages in exploration and evaluation. It is often underestimated, on the assumption that collecting a large volume of data is enough. In practice, however, what matters is not only the amount of information, but how that information is understood and correlated.

Interpretation is always linked to professional experience. The same dataset may lead to very different conclusions depending on who is analysing it. An experienced specialist may recognize patterns that are not obvious at first sight, correlate scattered observations correctly, and distinguish persistent features of a mineralizing system from incidental or misleading signs.

Interpretation, therefore, is not simply a description of the available data. It is an attempt to explain those data in terms of the geological processes and genetic setting that may have led to the formation of mineralization. It takes into account deposit type, geological setting, rock structure, and the possible mechanisms by which the mineralization formed.

As more data are collected and the interpretation is refined, a more integrated picture begins to appear. The position of mineralized zones becomes clearer, along with their orientation, possible extent, and internal variability.

At that point, individual observations cease to be seen as disconnected facts. They begin to form a system that can be described and tested.

This is where the need arises to move from individual data points to a geological model of the object.

The next stage of exploration and evaluation is no longer limited to correlating observations. It involves interpretation in the fuller sense: building a spatial representation of mineralization, a model that makes it possible to trace its distribution between points of observation and to assess its possible scale.

1.2.2 Geological Model and Its Limitations

Once individual observations begin to form a coherent picture, the next step is to move from a collection of data to a geological model.

A geological model is a spatial representation of the object. It reflects the interpreted distribution of rock units, structures, and mineralization between points of observation.

It is important to understand that a model is never an exact representation of reality. It is only an interpretation built on a limited amount of data.

In the subsurface, we are always dealing with isolated points of information: samples, drilling intervals, surface observations, and geophysical responses. Everything between those points is, inevitably, a matter of interpretation.

For that reason, every geological model contains uncertainty.

There are many different ways to build models. Different approaches are used for interpreting geological boundaries, defining mineralization domains, building surfaces, and interpolating grades. A great deal has been written on these methods, and no doubt much more will be written in the future.

For the purposes of project evaluation, however, what matters is not the variety of modelling methods, but the nature of the model itself.

A model is a tool.

It helps us visualize the spatial distribution of mineralization, reconcile data from different sources, and create a basis for further calculations. But it never fully coincides with the actual structure of the object.

In engineering, one simple idea is often repeated:

> All models are wrong, but some are useful.

That is especially true of geological models.

The appropriate level of model detail is not determined solely by the amount of available data or by the stage of the project. To a large extent, it depends on the nature of the mineralization itself.

Some deposit types are, by nature, relatively uniform and laterally persistent. Large massive bodies or stratiform systems with high geological continuity, for example, may show comparatively limited variability in both grade and geometry.

In such cases, excessive model detail often adds little value. The mineralization may be described adequately using relatively simple parameters: average grade, volume, and the overall geometry of the body.

An attempt to build a highly complex and detailed model for such objects may prove not only unnecessary, but harmful. Where data are limited, additional model complexity can create a false sense of precision and may lead to an accumulation of interpretive errors.

Overly detailed models also consume substantial time and money.

For that reason, the key issue is not the pursuit of maximum detail, but an understanding of what level of modelling actually matches the geological nature of the object.

A good geological model is not necessarily the most complex model.

A good model is one that describes the object adequately and serves as a practical tool for evaluation.

That is why one principle should always be kept in mind when building a model: the purpose of modelling is not to create the most elaborate description possible, but to produce a model that genuinely helps us understand the object and make decisions.

In current industry practice, the result of interpreting spatial geological data is most often a block model.

A block model is a mathematical representation of mineralization in space. The geological space is divided into blocks, and for each block values are estimated or interpolated for grade, density, and other parameters. In that way, the model provides a quantitative description of mineralization, both in terms of quality, meaning grade, and quantity, meaning volume.

The methods used to build such models may vary considerably. There are different approaches to domaining, grade interpolation, statistical analysis, and non-statistical modelling. A great many specialist books and technical guidelines have been written on these topics, and it makes little sense to examine those methods in detail in this book. Any reader interested in the technical side of modelling will easily find abundant open sources on the subject.

The key point is this: the interpretation of geological data ultimately produces a mathematical model that describes the spatial distribution of mineralization, its quantity, its grade, and its interpreted continuity.

In practice, that model is most often implemented as a block model.

A block model divides geological space into individual cells, or blocks. Each block has its own spatial parameters: its coordinates within the model and its dimensions. Taken together, these blocks form a three-dimensional grid within which mineralization is described.

A block, however, is not just a geometric cell. Its main role is that it acts as a carrier of information.

Each block contains attributes, meaning parameters that describe the properties of the rock and mineralization within that volume of space. At the geological interpretation stage, those attributes usually include geological characteristics such as rock type, domain assignment, grade, density, and other parameters derived from sampling data and geological interpretation.

It is important to understand that the set of model attributes should not be defined arbitrarily.

There is sometimes a temptation to include as many parameters as possible simply because someone has suggested that such attributes might be useful. In practice, that rarely adds value. Too many parameters make the model more complicated without adding real information.

In addition, the properties of mineralization and geological bodies are often relatively stable. Differences between individual blocks may be minor, and excessive attribute detail merely creates an illusion of precision.

A sensible approach, therefore, is to include only those attributes that are genuinely needed to understand the object and support subsequent calculations.

If additional parameters are needed later, they can almost always be added at that stage. New attributes can be calculated or interpolated on the basis of the existing data and the geological model.

For that reason, a block model should not be treated as a maximally detailed database. It should be treated as a working tool that contains exactly as much information as is needed for analysis and decision-making.

A block model should be as simple as possible, but still sufficiently informative to describe the mineralization and support further evaluation.

From a practical point of view, such a model can be thought of as a data table. Each block corresponds to a certain volume of rock in the ground and contains a set of parameters describing that volume: coordinates, block dimensions, geological characteristics, and grade.

It is from these blocks that tonnage, contained metal, and ultimately the economic value of the deposit are derived. All subsequent evaluation of the object, from resource estimation to economic assessment, is based on the information recorded in these blocks.

Consider a simple example.

For illustration, let us take a highly simplified case and assume that a block is completely filled with mineralized material. At this stage, we are not considering such factors as ore fraction within the block, waste fraction, mining losses, dilution, or other parameters that would be included in a more detailed evaluation. The purpose here is only to show the calculation principle.

Assume a single model block with dimensions of

$$20 \times 20 \times 10 \text{ m}^3.$$

The volume of such a block is:

$$20 \times 20 \times 10 = 4{,}000 \text{ m}^3.$$

For simplicity, let us assume an average rock density of 3.0 t/m^3.
The mass of material in the block is then:

$$4{,}000 \times 3.0 = 12{,}000 \text{ tonnes of material.}$$

Now assume that the average copper grade in this block is 0.8%.
This means that the quantity of copper in the block is:

$$12{,}000 \times 0.008 = 96 \text{ tonnes of copper.}$$

Thus, even a single model block already allows mineralization to be described quantitatively: the volume of material, its mass, and the amount of contained metal can all be determined.

It is important to note that, at this stage, we are speaking only about the mass of material. Whether that material is classified as ore or waste is determined later, once economic criteria such as cut-off grade are applied.

In a real block model, there may be thousands or millions of such blocks. The principle, however, remains exactly the same: each block describes a small volume of space through its block coordinates, X, Y, and Z, and contains the parameters from which tonnage, contained metal, and the economic value of the object are later derived.

1.3 FROM INTERPRETATION (MODEL) TO MINERAL RESOURCE

1.3.1 Quantitative Assessment of Mineralization

Once the geological model has been built, the next natural question is: How much mineralized material does the object actually contain?

Up to this point, the discussion has focused mainly on geological interpretation. The position of mineralized zones, their orientation, possible extent, and geological continuity have been defined. For further analysis, however, that is not enough. The mineralization must also be assessed quantitatively.

In other words, two simple questions must be answered:

- how much material is present in the ground
- and what grade does that material carry?

That is precisely what is meant by the quantitative assessment of mineralization.

The basis for such an assessment is provided by drilling data, sampling results, and the geological model. In practice, the calculations are usually carried out on the basis of a block model, in which each block contains grade, density, and other relevant parameters.

Once those blocks are combined in space, it becomes possible to calculate the total volume of mineralized material, its mass, and the quantity of contained metal.

In practical terms, the assessment of mineralization is essentially the aggregation of the information stored in the model blocks.

It is important to understand, however, that the values assigned to blocks are never direct measurements. Samples provide information only at discrete points in space. The values between those points are estimated using various interpolation methods and geological interpretation.

For that reason, any quantitative assessment of mineralization always contains a degree of uncertainty.

There are many methods available for carrying out such assessments. They may differ substantially in mathematical complexity and in the way data are interpolated.

A detailed discussion of those methods lies outside the scope of this book and belongs in a more specialized technical treatment.

For understanding the logic of deposit evaluation, however, the key issue is not the calculation method itself, but the result.

The result of such an assessment is a quantitative description of the object: the volume of mineralized material within the area under consideration, its average grade, and the amount of contained metal become known.

In different countries and under different reporting systems, such estimates may be classified in different ways. There are international reporting standards as well as national systems for the classification of resources and reserves. They use different terms and categories, but all of them are based on the same underlying task: to describe the quantity of mineralized material, its grade, and the level of confidence in that estimate.

For the purposes of further analysis, the most important outputs are the quantitative indicators themselves: material tonnage, grade, and total contained metal.

These are the parameters that are later used to evaluate whether development of the deposit may be economically justified.

1.3.2 Mineral Resources and Their Economic Meaning

At this point, mineralization can already be described quantitatively. The volume of mineralized material is known, its average grade can be estimated, and the approximate quantity of the valuable component can be calculated. At first glance, it may seem that this is the stage at which the object becomes a deposit. In reality, however, this is only the first step. A mineral resource is, first and foremost, a geological estimate of mineralized material based on the available data and their interpretation. It shows that a certain quantity of material is present in the ground, with a given grade and an interpreted degree of geological continuity. But such an estimate, by itself, does not yet show whether that material can be extracted at an economic profit.

At the resource stage, most of the factors that will later determine project economics are still unknown, or known only in approximate terms. The future mining method has not yet been defined. Processing or upgrading parameters are not yet established. Real recovery factors, likely mining losses and dilution, capital and operating costs, infrastructure constraints, and many other parameters are still uncertain.

For that reason, a mineral resource is, in essence, a geological characterization of the object.

It shows that the mineralization has sufficient scale and continuity to be regarded as a potential development opportunity. But it does not yet confirm the economic viability of the project.

That is why so-called modifying factors are generally not applied at the resource estimation stage. These factors, taken together, represent the technical, economic, metallurgical, legal, environmental, and other conditions that determine whether mineralized material can in fact be extracted.

The reason is not only that many of these factors are still unknown at an early project stage.

There is a more fundamental reason as well:

- A mineral resource describes the mineralization itself: its quantity, grade, and spatial distribution. In essence, it is a geological representation of the object.
- Modifying factors belong to a different system altogether, namely the mining and processing system. They depend on the selected mining method, equipment parameters, process flowsheet, economic assumptions, and many other variables.

If such factors are introduced at the resource modelling stage, the model effectively begins to incorporate dynamic coefficients that are not geological in nature but depend on assumed engineering decisions.

As a result, the original picture of the mineralization begins to be distorted.

The model ceases to describe the actual distribution of mineralized material and begins instead to reflect assumptions about future extraction. From the standpoint of geological evaluation, that is methodologically unsound.

For that reason, at the resource stage the model should describe the mineralization itself, as it is understood from the primary data.

The question of what proportion of that material may ultimately be extracted economically arises later.

In practice, this creates a certain paradox.

On the one hand, mineral resources are an essential foundation for further project development. Without a quantitative estimate, it is impossible to plan mining, design an operation, or carry out economic analysis.

On the other hand, the level of uncertainty at the resource stage remains high. Many of the parameters that determine the economic performance of the project are still unknown.

That is why project value at the resource stage usually remains relatively low.

The market understands that a resource is only a potential opportunity. To turn it into a real mining project, further exploration, engineering studies, metallurgical testwork, design work, and economic evaluation are required. All of this takes time and substantial investment.

Even so, many projects attract financing or change ownership at exactly this stage.

Investors and mining companies are often assessing not the current value of the resource itself, but its development potential. If the geological model is compelling, the scale of mineralization is significant, and the grade is consistent with expectations for that deposit type, the resource may provide a sufficient basis for further investment.

The next step is to determine what portion of the resource can in fact be extracted economically.

To do that, modifying factors must be taken into account: mining conditions, processing or upgrading parameters, economic assumptions, infrastructure, legal requirements, environmental constraints, and other conditions that may affect project implementation.

It is only after these factors are applied that a new category appears: reserves.

If a resource demonstrates the presence of mineralized material, reserves show what part of that material can be mined and used at an economic profit.

It is at that stage that the geological object is finally transformed into a mining project.

Before discussing reserves, mining, and mine design, however, a more fundamental question must be answered.

Mineralized material may occur in the ground in very large quantities. But volume alone means nothing.

The key question is always the same: under what conditions does that material begin to acquire economic value?

That is the point at which geological information begins to be viewed through an economic lens.

And this is where the next part begins: understanding how geology is turned into value.

2 The Economics of a Mineral Deposit

2.1 FROM RESOURCE TO SALEABLE PRODUCT

2.1.1 A Deposit as a Source of Future Cash Flow

One of the most expensive illusions in mining is to confuse the geological presence of a mineral commodity with the economic value of a project. After a successful drilling programme, the temptation is almost inevitable: the object has been found, volumes have been outlined, quality looks encouraging, and value therefore seems almost self-evident. But that is only the beginning. Between a deposit and a business lies a long, capital-intensive, and risky chain of decisions. It is along that chain that geology either turns into money or remains geology.

A geologist looks at an object through the shape of the orebody, thickness, raw material quality, variability, depth, structural features, and level of geological confidence. An investor, lender, or project executive looks at it differently. For them, the main question is not: how much mineral material is present in the ground? The real question is tougher: how much value can this object create, under what conditions, with what level of risk, and over what timeframe? That shift in the question opens the whole subject of this second part of the book. In mining, value is not discovered. It is designed, demonstrated, and then extracted through a system of technical and commercial decisions.

Nature does not create cash flow. It creates only the starting material for it. In one case, this may be copper, iron ore, or gold mineralization; in another, a coal seam; in a third, limestone, phosphates, graphite, potash, kaolin, industrial sands, crushed stone feed, or some other industrial mineral. But a natural occurrence is not, by itself, an economic asset. It becomes one only when it is clear how it will be mined, what it will become after extraction, whether the future product will meet market requirements, and whether enough money will remain after all costs to justify the capital invested.

This leads to an important principle. A deposit does not have one single built-in value. It has a range of possible values. The same geological object may become a robust, high-margin operation, a borderline project with fragile economics, or a resource that looks impressive in reports but performs poorly under investment scrutiny. Geology defines the boundaries of what is possible. But the final value is determined by the mining method, the process flowsheet, product quality requirements, logistics, market conditions, capital intensity, operating discipline, and quality of execution.

That is why value in mining emerges through a chain of conversion. Mineral material in the ground must first become the extractable part of the deposit. Then it must become material that will actually be mined and prepared. Then it must

DOI: 10.1201/9781042018932-2

become a saleable product that meets the required specification. Only after that do revenue and cash flow appear.

Deposit → extractable volume → mined material
→ saleable product → revenue → cash flow

At every transition, the system loses part of the apparent wealth that seemed obvious at the early stage. Some material remains outside pit limits or outside acceptable underground mining conditions. Some is diluted by waste and lost through mining and process inefficiencies. Some value disappears during processing, sorting, beneficiation, washing, drying, crushing, sizing, calcination, or other stages of product preparation. Some is absorbed by transport, energy, reagents, dewatering, stripping, infrastructure, treatment, impurity penalties, moisture losses, and other commercial terms of supply. That is why mining is not a business about raw material in the ground. It is a business about the ability to convert natural material, step by step and at a profit, into a product the market is genuinely willing to pay for.

This is where an important distinction appears:

- In metal deposits, attention is often focused on grade and recovery.
- In coal and industrial minerals, however, the logic of value shows even more clearly how many layers it contains.

The market does not buy just a tonne of mass taken from the ground. It buys a tonne of product with the required properties. In coal, those may include ash, moisture, volatile matter, sulphur, calorific value, coking properties, particle size distribution, and reliability of supply. In industrial minerals, the key parameters may be chemical composition, whiteness, purity, reactivity, hardness, particle size, deleterious impurities, homogeneity, and suitability for a particular application. In construction materials, value depends on strength, size fraction, frost resistance, undesirable components, proximity to market, and delivered cost. In all these cases, tonnage alone guarantees nothing. Value is created only by the tonnage that can be brought to market specification and sold at a profit.

This is one reason why projects are so often overvalued at early stages. The market likes simple stories: a large resource, good grades, straightforward geology. But a simple story almost never captures the full economics. A large object may still turn out to be a weak business if the product falls short of market requirements, requires expensive upgrading, does not travel well, is exposed to seasonal constraints, competes with lower-cost alternatives, or depends on one or two buyers.

Conversely, what looks at first like a modest deposit may prove to be an excellent asset if its material is easy to mine, simple to process, consistently capable of meeting specification, and saleable into a stable market with good logistics. In mining economics, the winner is not the object with more material in the ground but the one that converts that material more effectively into cash results.

This leads to a more precise value formula. Value is not simply volume multiplied by quality. Value is the volume of saleable product generated at an adequate profit and within the required timeframe. Both qualifications matter. Profit shows how much value actually remains in the project after the full production chain. Time shows what that value is worth today.

The time factor is especially important in mining. The same volume of product, produced at the beginning of the life of mine or near its end, may be physically identical but economically very different. Early cash flow returns capital faster, lowers risk, improves debt service capacity, and makes the project more resilient to adverse price conditions. Late cash flow depends more heavily on future markets, operating stability, infrastructure condition, regulation, and cost discipline. For that reason, the value of a deposit is determined not only by what it contains but by when the project can convert that material into saleable product and revenue.

A further fundamental feature must be added: almost every deposit is internally heterogeneous. Different parts of the deposit contain different raw material quality, different processing complexity, different moisture conditions, different mining depth, different accessibility, different metallurgical or technological behaviour, and different commercial value. This means that value is determined not only by what the deposit contains, but also by the sequence in which it is exposed and mined. Sequencing becomes an economic tool. What enters the first years of operation defines the project in the eyes of investors, banks, and owners. In some cases, early access to the best material is what makes the project financeable. In others, the absence of such access weakens even a very strong resource base.

In that sense, the boundary between useful and non-useful material is almost never purely geological. It is economic and technological. The same material may be valuable in one price environment, under one process route, and under one set of product requirements, and lose that value when those conditions change.

This is especially clear in iron ore concentrates, coal, and industrial minerals, where the decisive factor is often not an exchange-based view of the commodity, but the specific requirements of the product and the customer. The market does not pay for a geological report. It pays for the delivery of material that meets specification, arrives on time, and maintains the required quality from deposit to customer.

That is how experienced practitioners look at a deposit: not as a set of drill holes, maps, sections, and tables, although nothing begins without them, but as a system that must convert natural potential into a durable cash outcome. Good geology makes such a system possible. But only good geology combined with a realistic mining method, an adequate process flowsheet, a clear cost structure, correct market positioning, and disciplined execution makes it valuable.

The first rule of value assessment is severe, but liberating:

> do not confuse the presence of a mineral commodity with the presence of value. An object in the ground is not yet a business. It is only a possibility of future cash flow. And that possibility still has to be demonstrated through mining, processing, product specification, market fit, time, and capital.

That is where any discussion of value in mining should begin. Not with an impressive resource table and not with large volumes in the ground, but with the sober recognition that a deposit is not a warehouse of ready-made value. It is a system that may create cash flow, or may fail to create it. The whole logic of further evaluation is devoted to learning how to distinguish one from the other.

2.1.2 Modifying Factors and Cut-Off Grade

After the previous chapter, the next question follows almost automatically. If a deposit is not a warehouse of ready-made value, but a source of future cash flow, then what part of the geological body can actually become a business? Where is the boundary between material that looks attractive in geological terms and material that can genuinely be mined, processed, and sold at a profit?

This is where one of the most important transitions in mining takes place: the transition from geological presence to economic extractability.

As long as the discussion remains purely geological, a deposit may appear as a continuous mineralized body, a coal seam, or a mass of raw material of a given composition. But the moment the discussion shifts to mining, processing, product specification, logistics, and realized price, that apparent continuity begins to break down. Not all material has the same value. Not all material is equally accessible. And not all material should be treated as economically mineable.

This is precisely the role of modifying factors.

Modifying factors are the set of technical, mining, metallurgical, economic, legal, environmental, social, infrastructure, marketing, and other practical considerations that determine whether mineralized material can actually be converted into a viable mining proposition. They are the bridge between geological knowledge and mineable value.

Without them, a geological model remains only a geological model.

Once modifying factors are introduced, the evaluation begins to reflect mining method, losses and dilution, process flowsheet, metallurgical recovery, product quality requirements, infrastructure, permitting constraints, environmental obligations, operating costs, capital intensity, commercial terms, and market conditions. Only then does it become possible to ask not merely what is present in the ground, but what part of it may realistically support extraction and sale.

That is why reserves do not arise directly from the ground. They arise from the ground after geology has been filtered through a system of assumptions.

This is a fundamental point, because it changes the meaning of reserves itself. Reserves are not simply what exists in the ground. They are the part of the studied material that, under stated technical and economic assumptions, can be regarded as mineable and economically significant. In that sense, reserves are not just a geological statement. They are also an engineering and economic judgement.

A stricter formulation is therefore also correct: reserves are reserves at the time they are estimated.

Not in general.
Not forever.
Not outside time.
At the time of estimation.

This is one of the most underestimated truths in mining economics. When a reserve figure appears in a report, it is often treated as though it were an inherent property of the deposit, as if nature had delivered it together with the orebody itself. In reality, it is a conclusion reached on a specific date, at a specific price, under a specific cost structure, using a specific mining and processing concept, and within a specific market context. Change those assumptions, and the reserve changes with them.

This matters because both sides of the economic equation are always moving. On the revenue side, prices fluctuate, contract terms change, payables, premiums, and penalties shift, transport terms move, and customer requirements evolve. On the cost side, energy, fuel, explosives, reagents, labour, maintenance, water, stripping, processing, environmental compliance, and sustaining capital all change over time. In international projects, exchange-rate exposure often adds another layer: revenue may be earned in one currency, while much of the cost base sits in another. Reserve, therefore, is never a permanent geological constant. It is always a function of market conditions, technology, and time.

This is where cut-off grade enters the picture in its practical form.

Cut-off grade is the best-known numerical expression of the economic boundary within a deposit. In metal mining, it is usually understood as the minimum grade at which a tonne of material still covers its relevant costs and makes a positive contribution to the project. But cut-off grade is not a natural property of the rock. It is not something lying in the ground next to the mineralization. It is an economic threshold created by price, cost, recovery, mining assumptions, and commercial terms.

That is why cut-off grade is so useful as a teaching concept. It shows, in a single number, that value is not created in one drill hole and not contained in geology alone. It emerges at the intersection of geology, mining, metallurgy, market conditions, and time.

For coal and industrial minerals, the same logic applies even if it is not always expressed through a single assay threshold. In coal, the economic boundary may depend not only on ash or moisture, but also on product yield after beneficiation, calorific value, sulphur, coking behaviour, sizing characteristics, and contract terms. In industrial minerals, the decisive factors may include chemistry, whiteness, purity, deleterious elements, particle size distribution, suitability for end use, and the yield of saleable product after processing. But the principle remains unchanged: material becomes economic only when it can be converted into a marketable product at an economic margin.

Even so, the mechanism is easiest to demonstrate through a simple cut-off grade example.

Assume an open pit copper project. Let the copper price be USD 8,500 per tonne. Let metallurgical recovery be 88%. Let payable metal be 96% of recovered metal. Let the total cost that must be covered by one tonne of ore be USD 28 per tonne. For simplicity, assume these costs already include mining, processing, and site overheads. The purpose of the example is not to reproduce a full project model, but to illustrate the principle.

Now consider what 1% copper in one tonne of ore actually means.

One percent copper in one tonne of ore is 10 kilograms of contained copper, or 0.01 tonnes. After applying metallurgical recovery and payable terms, the payable copper per tonne of ore is

$$0.01\,\text{t} \times 0.88 \times 0.96 = 0.008448\,\text{t payable copper per tonne of ore}$$

Now multiply that by price:

$$0.008448\,\text{t Cu / t ore} \times \text{USD}\,8{,}500\,/\,\text{t Cu} = \text{USD}\,71.8 \text{ per tonne of ore}$$

Under these assumptions, each 1% copper in the ore generates approximately USD 71.8 of revenue per tonne of ore.

The cut-off grade is then found by asking what grade is required for one tonne of ore to cover USD 28 of cost:

$$\text{USD}\,28\,/\,\text{USD}\,71.8 = 0.39\%\ \text{Cu}$$

So, under these assumptions, material grading about 0.39% copper sits at the economic margin. Material meaningfully above that level contributes positive value. Material meaningfully below it destroys value.

But the real significance of the example appears when the assumptions change.

Suppose now that the copper price falls to USD 7,200 per tonne and the cost rises to USD 32 per tonne of ore. Recovery and payable terms remain unchanged. In that case, 1% copper now generates

$$0.01\,\text{t} \times 0.88 \times 0.96 \times \text{USD}\,7{,}200\,/\,\text{t Cu} = \text{USD}\,60.8 \text{ per tonne of ore}$$

The revised cut-off grade becomes

$$\text{USD}\ 32\,/\,\text{USD}\ 60.8 = 0.53\%\ \text{Cu}$$

The geology has not changed. The drill holes are the same. The mineralized body is the same. But the economic boundary of the deposit has shifted from 0.39% to 0.53% copper. And with it, the part of the model that can be treated as reserve has shifted as well. Blocks that looked economic yesterday may now fall outside the economically mineable part of the deposit. And the reverse is equally true: if price rises, costs fall, or recovery improves, some of that material may re-enter the reserve case.

This is the central point that is too often missed when reserves are spoken of as though they were fixed. Reserves are not a permanent attribute of the deposit. They are the outcome of a particular technical and economic configuration.

In real projects, of course, the calculation is more complex. It may include mining losses, dilution, variable recoveries by domain, haulage differences, royalties, treatment terms, penalties, product blending, multiple process routes, changing cut-off strategies, and, in more advanced studies, sequencing as well. But even in a simplified form, the lesson is clear: cut-off grade is not a geological constant. It is the numerical expression of the active modifying factors.

In its simplest form, cut-off grade can be expressed as

$$\text{Cut-off grade} = \text{Cost per tonne of material}\,/\left(\text{Product price} \times \text{Overall recovery}\right)$$

where:

Cost per tonne of material is the relevant cost that must be covered by one tonne of mined material.

Product price is the effective realized selling price after commercial adjustments.

Overall recovery is the total fraction of the valuable component that survives the full chain from in-situ material to payable product, including mining recovery effects, metallurgical recovery, dilution effects where relevant, and payable terms.

In this form, the equation makes the logic immediately visible. A change in any of the main variables, price, cost, or recovery, shifts the economic boundary of the deposit.

That is why it is impossible to talk meaningfully about reserves without talking about modifying factors. Two companies may look at exactly the same deposit and produce different reserve shells, different extractable tonnages, and different project valuations not because one of them "got the geology wrong," but because they are using different mining assumptions, different processing routes, different cost structures, different price views, different logistics, and different market assumptions. The geology may remain unchanged. What changes is the economic lens through which that geology is converted into a mining case.

In coal and industrial minerals, this point is often even more important. In metal mining, the threshold can often be approximated through grade and recovery. In coal, the practical question is usually different: which run-of-mine material, after mining and beneficiation, produces a saleable product that meets customer specification and covers its full cost? In industrial minerals, the same logic becomes sharper still: which material, after washing, sizing, purification, classification, drying, or other upgrading steps, still remains saleable and margin-bearing?

In such cases, the economic boundary is often defined not by a single number, but by a combination of quality indicators, saleable yield, and customer requirements. But the underlying principle remains unchanged. Modifying factors are the means by which market reality and process reality are imposed on geology.

An important practical consequence follows. Reserve figures should never be read as though they exist outside time. Whenever a reserve is quoted, the unspoken second half of the sentence should always be: reserve at what price, at what cost, under what mining and processing assumptions, in what market environment, and as of what date? Without that context, the number may look authoritative, but it says less than it appears to say.

This is the discipline of sound deposit evaluation. It does not merely total tonnages. It keeps in view the fact that the economic value of a deposit is created through a changing set of assumptions. Geology provides the material. Modifying factors determine which part of that material may become economically mineable. Cut-off grade gives that boundary a numerical form.

That is why, in mining, the question "what counts as ore?" is never purely geological. It is always a question of price, cost, recovery, mining method, process route, market, and time. And it is precisely at this point that a deposit begins to turn from a geological object into a project.

2.1.3 Why Grade Does Not Equal Value

In mining, one of the most common ways to describe a deposit is by reference to the grade of the valuable component. Reports, presentations, and company announcements usually speak in terms of metal percentages, grams per tonne, ash content, calorific value, chemical composition, or the concentration of a mineral in the rock.

That naturally creates a simple impression: the higher the grade, the more valuable the object.

In practice, however, the picture is far more complex. Grade alone does not determine the value of a deposit. It describes only the presence of a valuable component in the rock. Value arises only when it becomes clear whether that material can be extracted efficiently, processed successfully, converted into a saleable product, and sold into the market with a positive economic outcome.

One reason is that mineralization is rarely metallurgically or technologically uniform. Even within the same deposit, different ore types may occur which belong, geologically, to the same system, but behave very differently in processing.

This is especially clear where a single deposit contains oxide, sulphide, and mixed ores. In a report, they may show similar grades of the valuable component, but that does not mean they carry the same industrial value.

One ore type may be amenable to a relatively simple and stable process route. Another may be more difficult and more expensive to treat, with higher losses. A third may derive its main processing problems precisely from the interaction of different mineralization styles. In such cases, geological grade remains only an initial descriptor. The real value is determined by what can actually be delivered at the end of the process chain.

That is why two materials with the same grade may generate completely different economic outcomes. One may process well and deliver stable recovery. Another may require a separate flowsheet, produce lower recovery, incur higher cost, or result in inconsistent product quality. Formally, the grade may be identical, but the industrial value of the two materials may be very different.

In some cases, processing becomes the decisive factor that determines the fate of a project. Not because the geology is poor, but because the mineralization is difficult from a treatment point of view. This may be related to the mode of occurrence of the valuable component, the degree of oxidation, ore texture, the presence of deleterious impurities, a high clay content, or other characteristics that make processing more expensive, less stable, or simply less effective. As a result, one deposit with lower grade but simple and low-cost processing may be significantly more valuable than another with higher grade but poorer recovery, a more complex flowsheet, and materially higher costs.

For that reason, it is misleading to compare deposits on grade alone. Average grade is only one of the parameters used to describe mineralization. It does not show how the material will behave in processing, what product can be produced, how stable recovery will be, whether separate process streams are needed for different ore types, what losses will arise between the ground and the saleable product, or what all of this will ultimately cost.

In coal and industrial minerals, this logic is even more direct. The market does not buy mass from the ground. It buys a product with defined properties. In coal, the important parameters include ash, moisture, calorific value, sulphur, particle size distribution, and consistency of quality. In industrial minerals, they include chemical composition, whiteness, deleterious impurities, reactivity, strength, particle size, and suitability for a specific application. Here again, it becomes obvious that the mere presence of a valuable component in the ground does not mean that value exists.

An important conclusion follows. Grade is only the starting point of evaluation. It may show that an object deserves attention, but it cannot by itself answer the question of value. To understand the economic potential of a deposit, a much wider set of factors must be considered: ore types, processing behaviour, recovery, saleable product yield, cost structure, and market requirements.

2.1.4 From Material to Saleable Product

Even once it becomes clear that grade by itself does not mean value, the next question still remains: what, exactly, does the market pay for in mining?

At first glance, the answer seems obvious. If a valuable component is present in the ground, then surely that is what creates value. In reality, however, the market almost never pays for the mere presence of a valuable component in the ground. It pays only for a saleable product that meets defined requirements for quality, composition, consistency, and form of delivery.

The market itself is far from uniform. In mining, there is no single universal sales model that applies equally to all commodities. Some products are sold into relatively transparent and liquid markets, where price formation is more visible and is tracked almost continuously by market participants. This is especially true for precious metals and many base metals, where exchange quotations, or price references close to them, play a central role. In other segments, the situation is different. There, the market is built mainly on direct negotiation, long-term contracts, a limited number of buyers, tight dependence on quality specifications, and the physical realities of logistics. This is often the case for coal, a large part of the industrial minerals sector, and many specialized raw materials.

There are also intermediate cases. For some commodities, price indices and reference markers exist and create the appearance of an almost exchange-style market, yet actual sales still take place through contractual relationships and negotiations over quality, discounts, premiums, moisture, impurities, sizing, transport, and delivery terms. In such cases, the quoted price on paper and the real transaction price may differ materially. This is particularly important in iron ore products, manganese products, parts of the ferroalloy raw materials market, and other materials where market logic lies somewhere between open index pricing and tightly negotiated contract markets.

That is why, in mining economics, it is never enough simply to say that a product "has a price." One must always ask what market is being referred to, how price is actually formed in that market, how liquid it is, how many real buyers it contains, how important long-term relationships are, how strong the seller's negotiating position is, and what deviations from the benchmark price arise because of quality, volume, logistics, and payment terms.

The same product may look attractive in an abstract market assessment, yet in a real commercial setting its sales channel may be narrow, highly dependent on a small number of counterparties, and extremely sensitive to any deviation from specification.

This is where one of the key breaks between geology and economics appears. Geology describes material in the ground. The market pays for a product brought to the required specification. Between those two points lies an entire chain of conversion, and at every stage part of the potential value may be lost.

Not all of the valuable component contained in mineralized material can be recovered. Not all mined material reaches processing in the same condition. Not everything that enters processing becomes a saleable product. And finally, not everything that reaches product stage is fully paid for by the market.

Part of the value may be lost through incomplete recovery. Part may be lost through impurities, moisture, quality deterioration, or failure to meet product specification. Part may be lost because the route from material to product turns out to be technologically difficult, expensive, or unstable.

This matters because technical producibility is not the same as economic producibility.

In many cases, a saleable product can be produced technically, but the cost of achieving it is too high. One material may require only a relatively simple process route. Another may require additional upgrading stages, split process streams, special reagents, a more complex flowsheet, tighter quality control, or separate treatment for different ore types. All of this increases cost, complicates operations, and reduces the project's ultimate value.

That is why value is created not by the mere presence of a valuable component in the ground, but by the project's ability to convert raw material, consistently and at acceptable cost, into a saleable product that the market actually wants.

For metal deposits, that product may be a concentrate, a metal, or another payable product. For coal, it may be a product with the required ash, moisture, calorific value, and other parameters. For industrial minerals, it may be a product that meets specifications for chemical composition, whiteness, particle size, impurity content, strength, or suitability for a particular use.

In every case, the principle is the same. The market does not pay for a geological model, for a drill intercept, or for the mere quantity of valuable component in itself. It pays only for the product that can be made at the required quality and delivered to the point of sale.

That is why the path from material to product is one of the central segments of mining economics. It is here that one begins to see what part of the geological potential can truly be converted into revenue, and what part will remain only potential.

2.1.5 What Does the Market Actually Pay for?

Once the discussion of a deposit reaches the market, it quickly becomes clear that there is no simple or uniform picture here either. The market is often spoken of as though it were a single transparent system in which every product has some obvious and easily observable price. In practice, that is almost never the case.

In mining, the market is not one thing. Different commodities are sold through different pricing mechanisms, different sales structures, different product requirements, and different buyer groups. In some segments, price formation is more visible and can be read relatively easily. In others, price is determined mainly through direct negotiation, long-term relationships, a limited circle of participants, and a tight link to product specification. In still other cases, indices and reference prices may exist on the surface, yet the actual transaction price is still determined by contract terms, product quality, logistics, volume, and the negotiating strength of the parties.

That is why one of the most common errors in project evaluation begins the moment a product is assigned some abstract "market price" without first asking the more important question: in which market will this product actually be sold, to whom, and on what terms?

The market does not pay for a valuable component in the abstract, and it does not pay for mass alone. It pays for a specific product that meets a specific specification and fits a specific sales channel. This means that the same material may be viewed very differently by different buyers. For one customer, it may be an acceptable product. For another, it may be a second-grade material. For a third, it may be of no interest at all, even if it appears geologically attractive.

A further complication is that the market does not consist only of "buyers" in the abstract sense. Between the project and final consumption there are often traders, intermediaries, processors, affiliated structures, industry groups, large corporations, and vertically integrated players. Each of them has its own interests, and those interests do not necessarily align with the producer's interest in obtaining the highest possible price.

A trader wants to buy the product in a way that preserves its own margin and its control over the sales channel. A large industrial consumer is not primarily interested in "paying a fair market price," but in securing stable supply on terms that are as favourable as possible to itself.

Vertically integrated structures may view price differently again. For them, price may not be an independent market variable at all, but only one internal element within a larger system in which profit is captured further downstream, in logistics, in processing, or in the trading arm. That is why the statement "the market will pay" almost always requires clarification: who exactly will pay, for what exactly, in what form, and what share of that notional market price the project will actually retain.

This is where the distinction between price on paper and price in a real transaction becomes critical. In a presentation, it is easy to insert an index, a quotation, or a benchmark for an analogous product. In live commercial reality, however, price almost always becomes burdened with deviations: discounts for quality, premiums for consistency, penalties for impurities, moisture adjustments, sizing requirements, volume constraints, transport basis, payment terms, packaging requirements, shipment schedules, and the supplier's reliability. As a result, the same "market price" may translate into very different realized revenue for different projects.

For that reason, it is not enough to say that a product "has a market." That says far too little. One must understand what kind of market it is, how liquid it is, how many real buyers it contains, how independent they are from one another, how strong the seller's negotiating position is, how standardized the product is, how sensitive the market is to deviations in quality, and how the final transaction price is actually formed.

This is especially important in mining because many projects begin with an overly linear assumption: if the valuable component has been demonstrated, the process route has been designed, and a saleable product can be produced, then the market will naturally validate its value. In practice, however, this is often precisely the stage at which the next break appears between technical reality and commercial reality. A product may be technically achievable, yet commercially vulnerable. It may exist as a possible saleable product, but could be marketable only through a narrow group of counterparties, under steep discounts, with high dependence on a single customer, or under conditions where a substantial part of the potential value is captured not by the producer, but by other participants in the chain.

That is why, in mining economics, the market should never be treated as an external constant. It is not merely the background against which price appears. It is an active environment of interests, negotiation, constraints, and margin allocation.

An important conclusion follows. Value is created not only in the ground and not only in the plant. It arises only when the project can convert material into a saleable product and place that product into a specific market on terms under which a meaningful share of the value created remains with the project itself.

This is the point at which it becomes clear that even the existence of a saleable product does not yet guarantee a cash result. Between product and money there is always commercial reality, and commercial reality is not always favourable to the producer.

2.2 REVENUE, COSTS AND UNIT ECONOMICS

2.2.1 Revenue is Not Cash Flow

Once it becomes clear that the material can, in principle, be converted into a saleable product, and that a market exists for that product, the next natural question is this: where, exactly, do the money flows appear in that chain?

At first glance, the answer seems simple. If the product has a price and there is a buyer for it, then the project must already be earning money. In reality, however, another important boundary appears here, one that is very often underestimated in mining.

Revenue does not mean cash flow.

That distinction seems obvious only until real project evaluation begins. In practice, this is exactly the point at which many projects look better on paper than they do in reality.

Revenue is the value of the product sold. It shows how much money the project should receive for delivered material at given prices, volumes, and sales terms. But revenue alone says nothing about how much value remains in the project after that product has been mined, processed, and brought to the point of sale.

That is why product price and revenue represent only the upper part of the economic picture.

Below them lies the part that, in reality, determines the fate of the project.

To generate revenue, the material must first be extracted from the ground. But already at that stage, project economics begin to depend sharply on the nature of the material itself and on the selected mining method.

Material in the ground may occur as hard rock, semi-competent rock, unconsolidated sediments, coal seams, water-bearing formations, or weakly cemented units. In some cases, drill-and-blast preparation is required; in others, it is not. In some cases, the key driver is stripping volume, depth, and haul distance. In others, it is ground stability, dewatering, ventilation, support, backfill, or other features of underground mining. Costs, therefore, do not exist independently of orebody geometry, geotechnical conditions, hydrogeology, and the chosen mining method.

That is why costs can never be discussed as though they were a standardized list of operations. Their structure is always determined by a specific combination of factors: what is in the ground, how it can be mined, and what the material must become after extraction.

After mining, the material must be moved to the point where it either becomes a saleable product or enters further preparation and processing. Here again, an important distinction appears. For some commodities, relatively simple steps such as crushing, sizing, washing, blending, or basic preparation may be sufficient. For others, more complex beneficiation or metallurgical treatment is required. Between product and financial outcome, therefore, there is never an abstract production process, but a very specific chain of technical decisions shaped by the nature of the material and the selected project concept.

Another layer of cost is added by logistics. Even once the product has been made, it still has to be delivered to the point where the market takes ownership. For some projects, this may be only a short haul to a nearby customer. For others, it means rail transport, transshipment, port infrastructure, ocean freight, stockpile management, moisture losses, seasonality constraints, and the full system of commercial delivery. That is why a saleable product and an actually paid-for product are not always identical in economic terms.

In addition, every project incurs costs that are not tied directly to a single tonne of material, but without which the operation cannot function at all: management, maintenance facilities, laboratories, power supply, water supply, environmental compliance, security, communications, infrastructure, technical services, and the whole support system required to keep production running. That is why, between revenue and financial outcome, there always stands an entire structure of operating and organizational cost.

This is where the central distinction emerges.

A project may show substantial revenue and still remain weak in cash terms if the cost of generating that product is too high. Superficially, such a project may look convincing: there is a resource, there is a saleable product, and there is a market price. But if a large share of the revenue is immediately absorbed by mining, preparation, processing, logistics, and support of the production chain, then the project's real economic value is far lower than it appears at first sight.

To keep this distinction from remaining abstract, it is useful to look at a simple numerical example.

Let us take a deliberately simplified case. Assume that 1 tonne of ore feed yields 0.42 tonnes of saleable concentrate after processing. Let the selling price of that concentrate be USD 95 per tonne of concentrate. Revenue per tonne of ore feed will then be

$$\begin{aligned}\text{Revenue} &= 0.42\,\text{t concentrate / t ore feed} \times \text{USD}\,95\,/\,\text{t concentrate}\\ &= \text{USD}\,39.9\,/\,\text{t ore feed}\end{aligned}$$

At first glance, this already looks reasonably attractive. There is a product, there is a price, and there is almost USD 40 per tonne of ore feed in revenue.

But now we must ask what it costs to achieve that result.

Assume mining cost is USD 14 per tonne of ore feed, processing cost is USD 18 per tonne of ore feed, and commercial plus logistics cost is USD 11 per tonne of concentrate. The commercial and logistics component, expressed per tonne of ore feed, will then be

$$\begin{aligned}\text{Commercial and logistics cost} &= 0.42\,\text{t concentrate / t ore feed} \times \text{USD}\ 11\,/\,\text{t concentrate}\\ &= \text{USD}\,4.62\,/\,\text{t ore feed}\end{aligned}$$

Now the total direct cost can be calculated:

$$\begin{aligned}\text{Direct cost} &= \text{USD}\ 14\,/\,\text{t ore feed} + \text{USD}\,18\,/\,\text{t ore feed} + \text{USD}\ 4.62\,/\,\text{t ore feed}\\ &= \text{USD}\ 36.62\,/\,\text{t ore feed}\end{aligned}$$

The operating margin is then

$$\begin{aligned}\text{Operating margin} &= \text{USD}\,39.9\,/\,\text{t ore feed} - \text{USD}\,36.62\,/\,\text{t ore feed}\\ &= \text{USD}\,3.28\,/\,\text{t ore feed}\end{aligned}$$

This is the point at which the critical boundary appears.

Revenue looks convincing at USD 39.9 per tonne of ore feed. But after mining, processing, and commercial expenses, only USD 3.28 per tonne of ore feed remains. And this is still not the project's final cash flow, because general and administrative costs, sustaining capital, taxes, financing, and other unavoidable real-project items have not yet been included.

Now imagine that the market weakens slightly and the concentrate price falls to USD 85 per tonne of concentrate. Revenue then becomes

$$\begin{aligned}\text{Revenue} &= 0.42\,\text{t concentrate / t ore feed} \times \text{USD}\,85\,/\,\text{t concentrate}\\ &= \text{USD}\,35.7\,/\,\text{t ore feed}\end{aligned}$$

At the same costs, the result changes materially:

$$\text{Operating margin} = \text{USD } 35.7 / \text{t ore feed} - \text{USD } 36.62 / \text{t ore feed}$$
$$= -\text{USD } 0.92 / \text{t ore feed}$$

This simple example shows very clearly how sensitive a project can be to the realized selling price of concentrate. A decline of only USD 10 per tonne of concentrate is enough to eliminate the positive operating margin and make the project uneconomic at this level of calculation.

In other words, even without a complex model, one can already see one of the key features of mining economics: a project may appear acceptable under one market scenario and lose its resilience very quickly under a weaker one.

That is why even very simple indicators can reveal the project's sensitivity to key parameters, in this case, the selling price of the product.

Exactly how to work with that sensitivity, how to evaluate project robustness, and which tools can be used to manage such risks will be discussed later.

This is why, in mining, one cannot stop at product price and not even at revenue. The real question is always different: how much money remains after the product has been produced, brought to market, and sold?

Every tonne of mined material, and every tonne of saleable product, matters only to the extent that something remains in the project after the associated costs have been paid. If the material can technically be mined and processed, but the cost of doing so absorbs almost all of the revenue, then its significance to the project is weak. If, after covering those costs, a sufficient amount remains, that remainder is what begins to create durable economic value.

This distinction matters for another reason as well: in mining, the movement of money almost never coincides in time with the movement of product.

First, the project spends money on stripping, mining, preparation, processing, transport, sales logistics, and maintaining the entire production system. Only after that does revenue appear. Cash flow is therefore sensitive not only to the level of cost, but also to when those costs are incurred and when payment for the product is actually received.

That is why even a project that looks profitable on paper may come under serious cash-flow pressure if costs arise earlier, faster, or in greater volume than revenue is collected.

An important conclusion follows. In mining economics, at least three levels must be distinguished: product price, sales revenue, and the cash flow that actually remains in the project.

- Price speaks about the market.
- Revenue speaks about the sale.
- Cash flow speaks about the viability of the business.

It is cash flow, not revenue, that ultimately determines whether a project can repay capital, service obligations, finance development, survive adverse market cycles, and exist at all as an industrial enterprise.

That is why, in a sound deposit evaluation, the question is never framed simply as "what is the product worth?" That question is too crude. The correct question is different:

> how much money remains in the project after that product has been produced, delivered, and sold?

This is the point at which revenue stops being a pleasing number in a table and becomes a real economic test of the project.

And that leads directly to the next logical step.

If revenue by itself does not mean cash flow, then the decisive issue is not only how much the market pays, but also how much the project spends to create and sell that product.

That is why the next task is to understand how costs determine whether value survives in the project, or disappears on the way from the ground to the sale.

2.2.2 Costs Determine Whether Value Survives

Once it becomes clear that revenue, by itself, does not mean cash flow, the next question follows almost inevitably: what, exactly, determines whether value remains in the project after the product is sold, or disappears on the way from the ground to the market?

The answer seems simple: costs.

But in mining, that simple word conceals a much more complex reality.

Project cost is not a single number and not a single metric. It is an entire system of interrelated parameters, each of which affects the project's final profitability. That is why the economics of a deposit cannot be understood as a linear equation in which one simply inserts price, volume, and grade and obtains a ready answer. In practice, it is much closer to a multi-parameter system that must first be assembled correctly, understood, broken down into its components, and bounded in terms of what each parameter really means before anyone can hope to solve it properly.

That is why a sound project evaluation does not begin by trying to "calculate the answer." It begins by understanding the structure of the system itself.

- Which parameters are primary.
- Which are secondary.
- Which are stable.
- Which are volatile.
- Which depend on geology.
- Which depend on processing or mining method.
- Which are driven by the market.
- And which are determined by engineering quality and operational execution.

Until that system has been unpacked properly, any final number may look neat, but say far too little.

In that sense, costs are not merely the expense side of a project. They are the structure through which value is either preserved or destroyed.

On one project, the problem may lie in mining. On another, in processing. On a third, in logistics. On a fourth, in a combination of several factors, each of which appears acceptable on its own, but together makes the project fragile or uneconomic.

That is why costs can never be treated as something homogeneous.

They always consist of different groups of parameters.

One group is determined by the nature of the deposit itself: depth, orebody geometry, ground conditions, the physical and mechanical properties of the material, hydrogeology, grade variability, and the distribution of ore types or quality classes of raw material.

Another group is determined by the selected mining method and process route. The same material can be mined and processed in different ways, and therefore the cost structure can vary greatly. In some cases, a project appears robust only under one specific development concept. In others, a potentially good deposit becomes economically weak simply because the wrong operating or processing scheme has been chosen.

There is also a third group of parameters: organizational and infrastructure factors. Equipment productivity, haul distance, seasonality, availability of water and power, distance to processing or market, infrastructure loading, consistency of product quality, and reliability of the supply chain all become part of the economic equation.

That is why two projects with very similar geology may generate very different profitability.

Even where grade, tonnage, and product appear comparable, value may survive in one project and be destroyed in another simply because the system parameters are different.

- In one case, the cost of extracting the material is lower.
- In another, processing is better.
- In a third, logistics are shorter or more efficient.
- In a fourth, product quality is more stable.
- In a fifth, losses and dilution are lower.

And all of these differences ultimately turn not into "technical details," but into differences in profitability.

That is why project profitability should never be seen as a passive result. It does not simply "come out of" the deposit. It is formed through a system of decisions and constraints.

This matters especially because many of the economic parameters in a mining project are highly sensitive even to small changes. Where a project operates with only a narrow margin of robustness, a change in a single parameter may be enough to eliminate the positive result.

- A small drop in recovery.
- A small increase in transport cost.
- A small deterioration in product quality.

- A small increase in losses.
- A small rise in unit processing cost.

Each of these changes may appear minor in isolation. But inside a multi-parameter system, such changes can shift the entire result.

That is why the task of a competent evaluator is not only to see the final number, but to understand which parameters create that number and which of them are critical.

This is one of the main differences between sound economic evaluation and superficial assessment. A superficial assessment tries to reach the final answer as quickly as possible. A good evaluation first tries to understand how the project equation is actually structured.

- Which parameters are imposed by the nature of the deposit.
- Which are engineering choices.
- Which are dictated by the market.
- Which can be improved.
- Which are barely controllable.
- And where the real levers of profitability lie.

At this point, another important idea begins to emerge, one that will run through the rest of the book.

Project profitability is not merely a characteristic. It is an object of management.

- It can be strengthened.
- It can be weakened.
- It can be destroyed by poor decisions.
- And it can be improved through disciplined work on the right system parameters.

But before that can be discussed seriously, the parameters themselves must be understood.

- Which of them have the strongest influence on the project's fate.
- Which are linked to one another.
- Which define the limits of what is possible.
- And which open room for improvement.

That is why the next step is not to "solve the whole equation" immediately, but to learn to see the project's sensitivity to its key parameters.

Only then does it become clear where profitability is actually created, where it is lost, and which decisions are genuinely capable of improving it.

2.2.3 Sensitivity: Which Parameters Matter Most?

Once it becomes clear that revenue, by itself, does not mean cash flow, and that costs determine whether value survives in a project or disappears along the way, the next question follows almost automatically: which parameters in that system are actually decisive?

At first glance, the answer may seem obvious. The more parameters are included, the more complete the evaluation must be. In real mining economics, however, completeness is only part of the task. It is just as important to understand which parameters actually move the result, and which change it only marginally.

That is the purpose of sensitivity analysis.

A project is almost always described by a large number of variables: product price, recovery, throughput, feed quality, losses, dilution, transport, moisture, impurity content, saleable yield, reagent consumption, energy use, fuel cost, stripping ratio, logistics, capital expenditure, quality consistency, payment terms, and many others. But not all of these variables influence profitability to the same degree.

In one project, the decisive parameter may be price. In another, recovery. In a third, saleable yield. In a fourth, logistics. In a fifth, deleterious impurities. Sometimes a project appears strong until it becomes clear that its resilience depends almost entirely on one or two vulnerable parameters. In other cases, a project may look modest, yet prove robust precisely because its economics are not critically exposed to small movements in the key drivers.

That is why the purpose of sensitivity analysis is not simply to list project parameters. Its real task is to identify which of them actually determine the fate of the project.

This is the point at which a project stops being just a calculation and starts to be understood as a system.

That matters because project parameters almost never exist in isolation. They are linked to one another, and a change in one often triggers changes in several others. For that reason, a good sensitivity analysis cannot be limited to checking one parameter at a time as though each of them lived separately from the rest of the system.

At first sight, for example, it may seem that increasing recovery always improves project economics. In reality, higher recovery often requires a change in process route, finer grinding, additional reagents, longer residence time, tighter quality control, or a more consistent and higher-quality feed. That, in turn, changes costs, throughput, sometimes product yield, and even the requirements imposed on mining itself.

In other words, a parameter cannot be considered in isolation.

An attempt to improve recovery may simultaneously:

- increase operating costs,
- reduce plant throughput,
- change feed quality requirements,
- alter the structure of the mined material,
- and affect overall profitability in different ways.

That is why a sound evaluation must look not only at the parameter itself, but also at the other parameters it pulls with it.

The same applies to other variables.

- Reducing losses may require more selective mining and increase mining cost. Reducing dilution may lower processing tonnage, but improve product quality.
- Improving product quality may require tighter material selection and reduce tonnage.
- Reducing logistics cost may require a different logistics chain or a different project scale.

In all such cases, a decision that is locally correct does not necessarily produce the best result for the system as a whole.

That is why sensitivity analysis is needed not only to determine which parameter has the strongest effect on a project. Its more important purpose is to show how the project responds to changes in its parameters, and where its real economic constraints lie.

In practice, this means that the evaluator must ask not one question, but several at once:

- Which parameter has the strongest influence on profitability?
- Within what range is its variation still acceptable?
- Which other parameters move together with it?
- Does a local optimization improve the overall result, or merely create the illusion of improvement?

And where are the points at which the system becomes unstable?

That is why project sensitivity should never be treated as a formal table or as a technical appendix to a calculation. In reality, it is one of the main tools for understanding the project.

It shows where the object has real weaknesses.

- Where the main risks lie.
- Which parameters deserve the greatest attention.
- And, most importantly, where the real levers for improving profitability are located.

Because project profitability is not just the output of a calculation. It is the result of how the entire system of parameters behaves.

At this stage, it is still too early to speak about solving the full system. First, its structure must be understood: which parameters are primary, which are derived, which are stable, which are volatile, which are barely controllable, and which, by contrast, can be improved through engineering, process, or organizational decisions.

That is where rational work on profitability begins.

Not with an attempt to produce a single final number as quickly as possible, but with an understanding of what actually determines that number.

That is why good project analysis always begins with the question: which parameters in this system matter most?

Because only then does it become possible to move to the next step and understand how those parameters change over time, and why the same project economics may look very different today and tomorrow.

2.3 TIME, SEQUENCE, AND CASH FLOW

2.3.1 Time Changes Value

Up to this point, the project has been viewed in a kind of conditional static state. Price, cost, recovery, quality, logistics, and other parameters were treated as given, and the task was to understand the economic result that follows from those assumptions.

In reality, however, a mining project never exists in a static state.

It evolves through time.

And as it evolves, the parameters on which its economic evaluation is built evolve with it.

- Prices change.
- Costs change.
- Market requirements change.
- Logistics change.
- The quality of mined material changes as extraction advances.
- Understanding of the deposit improves.
- Process routes change.
- Metallurgical assumptions change.

That is why project economics can never be fixed once and for all. Every evaluation is tied to a particular moment in time and to the assumptions adopted at that moment.

This is one of the defining features of mining economics.

The same deposit may look robust and profitable under one market environment, marginal under another, and unable to pass economic scrutiny under a third. The geological foundation of the object may remain unchanged. What changes is not the deposit itself, but the economic lens through which it is being evaluated.

This is seen especially clearly in cut-off grade.

Earlier, it was already shown that cut-off grade is not a geological constant. It depends on price, cost, recovery, and other project conditions. As those parameters change, the boundary between material that can be treated as economic and material that falls outside economic justification changes with them.

That is why time, in mining economics, is not simply a calendar.

It is a factor that changes the parameters of evaluation.

What appears to be economic material under one price and one cost structure may cease to be economic under another. And the reverse is equally true: material that lies outside the economic boundary today may re-enter the calculation if market conditions change.

This means that project evaluation itself must not be treated as a final verdict. It is a result for a specific set of assumptions at a specific point in time.

Of all the parameters that change over time, price is usually the most sensitive and at the same time the least controllable.

Price affects revenue, cut-off grade, the economic envelope of the project, and its overall resilience more strongly than almost any other variable. Yet price is precisely the parameter that cannot be known in advance with sufficient certainty.

- A project can be better drilled out.
- A process route can be improved.
- Logistics can be optimized.
- Losses and dilution may be reduced.

But future price can only be forecast.

This creates one of the hardest tasks in any serious project evaluation.

A project must be assessed using a future price that cannot be known exactly. The task, therefore, is not to "guess the market," but to choose assumptions that can be justified and that do not distort the project either through artificial optimism or through unnecessary understatement of its potential.

This is a fundamental point.

Deliberately inflated price assumptions may make a project look attractive on paper, but only on paper. They may expand the apparent economic envelope, reduce the visible quality threshold, create the illusion of stable profitability, and push decision-makers toward choices that will later prove wrong in the real market. In mining, the cost of such an error is particularly high, because decisions are usually tied to large capital commitments, long development cycles, and hard-to-reverse execution paths.

But artificially low assumptions are dangerous as well.

If the price assumption is set too low without sound justification, the project may be distorted in the opposite direction. The economic envelope may be compressed artificially, potentially mineable material may fall out of the evaluation, and a good project may appear weaker than it really is. The result may be rejection of an object that would have been viable under realistic assumptions, or underestimation of its importance at a stage when financing could still have been attracted and the project advanced.

That is why price forecasts must be treated seriously.

- Not as a formality.
- Not as a number that can be adjusted to fit a desired conclusion.
- And not as a place for excessive conservatism.

Price is one of the most important, and one of the most dangerous, parameters in the entire system.

But it is not the only parameter that changes with time.

Costs do not remain constant either. Fuel, energy, reagents, labour, transport, equipment, maintenance, infrastructure, and contractor services all move over time. In international projects, exchange-rate exposure is often added to the picture: the product is sold in one currency, while a large portion of the cost base sits in another.

As a result, even with a stable product price, project economics may change materially because the cost side is moving.

The project itself also changes over time.

- As work advances, the geological model may be refined.
- Assumptions about losses and dilution may change.
- The process route may be improved, or, in some cases, may prove more complex than expected.
- Understanding of the quality of different zones and material types may evolve.

In other words, time affects a project not only from the outside, through the market, but also from the inside, through the accumulation of knowledge and the revision of engineering decisions.

That is why project evaluation must always remain open to revision.

This is not a weakness of evaluation. It is not a flaw in the analysis. It is the normal consequence of the fact that a mining project exists inside an evolving system of parameters.

A sound evaluation does not try to conceal that variability. On the contrary, it recognizes it and tries to understand which parameters move the result most strongly over time.

This is one of the main bridges between static calculation and real project economics.

As long as parameters are treated as fixed, the project looks like a problem with one answer. But as soon as it becomes clear that price, cost, quality, recovery, and other factors change, it becomes obvious that we are no longer dealing with one answer but with a range of possible project states.

This is where analytical discipline becomes especially important.

One must understand:

- Which assumptions have been adopted.
- Why those assumptions have been chosen.
- Within what range they may change.
- What consequences those changes have for profitability.
- And at what point the project ceases to be robust.

Only under that approach does time-related variability stop being a source of chaos and begin to become part of disciplined evaluation.

And from here a very practical question arises.

If project parameters change through time, and if the economic boundary is not fixed, then the sequence in which material is extracted cannot be economically neutral. In other words, mining sequence begins to affect not only the schedule, but the economics of the project itself.

That is why the next step is to understand that sequencing is not just a technical timetable. It is a full economic instrument.

To keep this point from remaining purely conceptual, it is useful to look at a harder example.

Assume that a copper deposit was evaluated and advanced during a period of strong market conditions. At the time of evaluation, the copper price was USD 12,000/t Cu, and the average grade within the economic shell was taken at 0.35% CuEq.

It is worth clarifying immediately what metal equivalent means. A metal equivalent grade is a converted grade in which several payable components are translated into the equivalent grade of one base metal. In other words, if the ore or concentrate contains copper together with gold, silver, or another payable component, the contribution of those additional components is converted into an equivalent copper grade using assumptions for price, recovery, and payability. This is a useful indicator for broad evaluation, but it always remains a calculated value that depends on the assumptions adopted.

Under those conditions, the project appeared reasonably robust, and the adopted assumptions supported treatment of the full material volume included in the economic shell.

But for a sulphide concentrate project, the copper market price alone is not enough. The project does not capture the full headline metal price. It captures only the portion that remains after deductions for metallurgy, commercial terms, and logistics. In addition, not all of the contained metal in the ore becomes revenue. Only the part that makes it all the way from the ground to payable concentrate enters the revenue line.

For simplicity, let us assume the following:

average grade: 0.35% CuEq, or 0.0035 t CuEq per tonne of material
metallurgical recovery to concentrate: 88%
payable share of metal in concentrate: 96%
overall recovery: $0.88 \times 0.96 = 0.8448$, or 84.48%
combined deductions for metallurgy, commercial terms, and logistics: USD 1,500 per tonne of payable Cu

At a copper price of USD 12,000/t Cu, the effective price working inside the project is therefore not USD 12,000, but

$$\text{USD } 12{,}000 \,/\, \text{t Cu} - \text{USD } 1{,}500 \,/\, \text{t Cu} = \text{USD } 10{,}500 \,/\, \text{t Cu}$$

The payable copper per tonne of material is

$$0.0035 \text{ t CuEq} \,/\, \text{t material} \times 0.8448 = 0.0029568 \text{ t Cu} \,/\, \text{t material}$$

Revenue per tonne of material is then

$$0.0029568 \text{ t Cu} \,/\, \text{t material} \times \text{USD } 10{,}500 \,/\, \text{t Cu} = \text{USD } 31.05 \,/\, \text{t material}$$

This is how price actually begins to work inside the project.

Now suppose that, after the project starts, market conditions deteriorate sharply and the copper price falls to USD 4,500/t Cu.

After the same deductions for metallurgy, commercial terms, and logistics, the effective project price becomes

$$\text{USD } 4{,}500 / \text{t Cu} - \text{USD } 1{,}500 / \text{t Cu} = \text{USD } 3{,}000 / \text{t Cu}$$

At the same overall recovery, revenue per tonne of material becomes

$$0.0029568 \text{ t Cu} / \text{t material} \times \text{USD } 3{,}000 / \text{t Cu} = \text{USD } 8.87 / \text{t material}$$

In other words, on a per-tonne basis, revenue falls from USD 31.05/t material to USD 8.87/t material

The difference is

$$\text{USD } 31.05 - \text{USD } 8.87 = \text{USD } 22.18 / \text{t material}$$

For a mining project, this is a massive change.

If the original project economics were built with only a limited margin of robustness, such a fall in price means that a substantial part of the material that was previously regarded as economic no longer passes the economic test. On paper, the deposit has not disappeared. But the operating project begins to live under a completely different set of rules.

This is the point at which there is often a practical temptation to shift into selective mining of the higher-grade areas.

In the short term, that can look like a reasonable response. If the same deposit contains higher-grade zones, the project begins to rely on them in order to preserve at least a minimally positive cash result. In other words, the best parts of the deposit start compensating for the weaker market environment.

But that decision has a second side.

First, selective mining almost always breaks the original logic of the project. It disrupts the balance on which assumptions about throughput, plant feed quality, mine life, and use of the mineral base were built.

Second, accelerated extraction of the best areas means that the project begins to consume its most valuable material earlier than planned, leaving weaker material for the future.

Third, this can materially weaken the project over the longer term. Over a short period, such a tactic may indeed support revenue and cash flow. But later the operation is left with a poorer, more complex, and less resilient remainder of the mineral base. The project that was being "saved" through selective mining may, in the longer horizon, be fundamentally damaged by it.

That is the main point of the example.

A change in a single parameter, price, can do more than simply weaken project economics. It can change the operating mode of the entire project. What looked, at the evaluation stage, like one coherent and robust project may, after a market deterioration, turn into an operation that survives only by accelerating extraction of its best material.

That is why time-related variability affects not only the final number in the model. It can change the way the project itself behaves.

And that is precisely why price assumptions must be treated with maximum seriousness. They must not be knowingly inflated. But artificially depressing them is dangerous as well, because in both directions the consequences may be severe.

2.3.2 Mining Sequence as an Economic Instrument

Once it becomes clear that project parameters change through time, the next question follows almost automatically: does the order in which material is extracted affect project economics?

The answer is yes.

And it does so not as a secondary production detail, but as a full economic instrument.

Mining sequence determines not only the order in which material is removed from the ground. It affects plant feed quality, early revenue, early cash flow, project resilience under adverse market conditions, the economically viable life of the operation, and ultimately the overall profitability of the project.

That is why mining sequence should never be treated as a neutral production schedule. It is part of the project's economic architecture.

An important qualification is needed immediately, however.

The subject of sequencing is often oversimplified. It is easy to create the impression that all one has to do is move parts of the deposit around in time, bring higher-grade material forward, raise average feed quality in the early years, and thereby improve the project automatically. Over a short horizon, that approach can indeed produce a visible effect. Higher plant feed quality at the beginning of the mine life can raise revenue, improve early cash flow, and make the project appear more robust.

But this is exactly where one of the most dangerous illusions in mining economics begins:

- Not every change in sequence is beneficial.
- And not every improvement in the early years improves the project as a whole.

If a project pulls too much of its best material into the front end of the mine life, it may achieve a local improvement in early economics while weakening the remaining part of the deposit. The later stages of the operation then have to run on weaker material, with poorer feed quality, lower resilience, and less ability to withstand weaker market conditions.

In other words, the project may become more attractive at the beginning and weaker in total.

This is especially dangerous because early improvement is very easy to mistake for success. Average grades rise, the first years produce stronger cash results, the model becomes more attractive, and the internal logic of the project appears to be validated. But if this is achieved by accelerating extraction of the most valuable part of the mineral base, then the decision may represent not an improvement of the project, but a transfer of its problems into the future.

That is why mining sequence must not become an object of mechanical rearrangement.

It is not a question of:

can the schedule be made to look better?

It is a question of:

how does a change in sequence affect the project as a whole?

That is the correct way to view it:

- If higher-grade zones are brought forward, what happens to plant feed quality in the later years?
- How does the length of economically robust operation change?
- Does the project become too dependent on the first years?
- Does the quality of the remaining mineral base deteriorate?
- Does the project become more sensitive to a future fall in price?
- Is a short-term gain being created at the cost of long-term weakening?

Until these questions are answered, any "optimization" of sequence remains too superficial.

That is why sequence decisions should not be made as local improvements to a production calendar, but as the result of a sensitivity analysis of the entire project in response to changes in the order of extraction.

This is a fundamental point.

If a project is highly sensitive to early plant feed quality, then sequencing may indeed be a powerful lever.

If the project is especially vulnerable to lower prices, early access to stronger material may provide an important buffer.

But if an attempt to improve the early years destroys the structure of later stages, shortens the economically viable life of the operation, or materially weakens late-stage economics, then the decision cannot be regarded as sound, even if it looks attractive on the first pages of the model.

In other words, mining sequence should not be judged by how effectively it improves the front end of the project, but by how it affects the entire project over its full life.

That is the mature view of sequencing.

- Not to "improve" the object at any price.
- Not to pull the best pieces out of the system for the sake of a local gain.
- Not to turn sequence into a cosmetic tool for the model.

But, instead, to understand that sequencing is one of the ways in which profitability, resilience, and the distribution of economic result through time can be managed.

That is why, before sequence is adjusted, the structure of the project itself must be understood:

- Where its real weaknesses lie.
- Which parameters are most sensitive.
- How strongly it depends on the quality of the early years.
- How robust the later stages are.
- And whether a local improvement destroys the long-term logic of the project.

Only after that analysis does sequence become a true management tool.

Without it, sequence becomes only a pleasing but dangerous illusion of control.

The central conclusion follows directly.

Mining sequence is a powerful economic lever. But using it without a system-level sensitivity analysis of the whole project is unsound. Otherwise, the early years may be improved at the cost of weakening the entire downstream economics of the operation.

And this leads directly to the next step.

If the order of extraction affects when the project receives its strongest and weakest cash results, then another fundamental issue must be understood: why early and late money are not equal in value.

2.3.3 Why Early Cash Flows Matter

Once it becomes clear that mining sequence affects feed quality, revenue, and project resilience, the next question follows naturally: why does it matter so much when those cash flows actually appear?

At first glance, it may seem that only the total result matters. If a project ultimately generates a certain amount of money, then the exact timing of that result should not matter too much. In mining, however, that view is almost always too crude.

For a project, it is not enough to ask how much money it creates. One must also ask when it begins to create it.

That is why early cash flow is almost always more valuable than late cash flow.

This is not merely a financial formality or an abstract theoretical point. It follows directly from the way any mining project is structured. Money is spent first. The deposit must be studied, prepared for development, connected to infrastructure, opened up for mining, and equipped for extraction, processing, logistics, and sales. Only after that does the project begin to return capital.

That is why early positive cash flow has special importance.

First, it returns invested capital faster. The earlier a project begins to generate real cash, the faster the pressure of the initial investment starts to ease.

Second, early cash flow reduces overall project risk. Mining operates in an environment where parameters rarely remain constant: prices change, costs change, material quality changes, market conditions change, logistics change, product requirements change, and the broader business environment changes with them.

Cash generated in the early years is exposed to fewer of these uncertainties than cash the project hopes to earn far in the future.

Third, early cash flow increases the resilience of the operation. It makes debt service easier, provides a financial buffer, helps the project survive weaker market conditions, and gives management more freedom in future decisions.

Fourth, early cash flow makes the project more manageable. When an operation begins to return cash relatively quickly, it gains the ability to finance flowsheet improvements, infrastructure upgrades, additional studies, plant expansions, or other measures that strengthen the business. When the economic result is pushed too far into the future, the project becomes much more tightly locked into its original assumptions and much less tolerant of errors in evaluation.

That is why two projects with the same total economic result may have very different economic value.

One project may return a substantial part of its cash in the first years and then operate from a position of greater stability. Another may deliver a similar cumulative result only over a long horizon, while remaining far more vulnerable throughout. Formally, total revenue or total cash generation may look similar. But from the standpoint of the investor, the lender, and the project itself, these are not equivalent situations.

The same is true even within the same deposit.

Even if the mineral base remains unchanged, and even if the total amount of potential cash generated does not change, a different mining sequence can make the project either stronger or more fragile. If the better zones, the more stable feed quality, or the more profitable material fall into the early years, the project gains an advantage. If, on the other hand, the strongest part of the deposit is pushed too far into the future, while the first years rely on weak economics, the project may look heavy and vulnerable even if the mineral base as a whole is sound.

In that sense, time becomes part of the economics itself, not an appendix to it.

Late cash flow is almost always weaker in economic significance. Not because the money is somehow "worse" in itself, but because the project must still survive long enough to reach it. By that time, market conditions may have changed, costs may have risen, feed quality may have deteriorated, technical problems may have appeared, logistics may have tightened, environmental requirements may have hardened, exchange rates may have moved, or other factors may have weakened the project. In other words, late cash is always more heavily burdened by uncertainty.

That is why mining economics cannot look only at the total amount of result. It must also look at how that result is distributed through time.

This is one reason why simply trying to maximize the total outcome is not always the right objective. In some cases, it is more important to make the project stronger in the early years than to increase its formal total result in a distant future. And the reverse is also true: a project that looks impressive in terms of total future cash may still be weak precisely because too much of that result lies too far out in time.

That is why early cash flow plays such an important role in the investment logic of a mining project.

It shows not only that the object is capable of earning money. It shows how quickly the project begins to convert invested capital into a real economic result, how resilient it is to adverse change, and how viable its overall structure really is.

This leads to the next principle.

Future money and money today are not the same thing.

That is why a project cannot be evaluated only by the sum of its expected future receipts. It is also necessary to take into account when those receipts actually arise.

2.3.4 The Risks of a Mechanical Approach to Project Evaluation

Once it becomes clear that early cash flow matters more than late cash flow, the next natural question is how this should be reflected in project evaluation in practice.

This is where discounting enters the picture.

In the broadest sense, its meaning is simple. Future money is not equal to money today. Cash that a project expects to receive several years from now has less value than cash received now. The reason is not time alone, but also risk, uncertainty, the cost of capital, possible market changes, and the fact that the project must still survive long enough to reach those future cash flows.

That is why future cash flows are brought back to a single reference point, usually the date of evaluation. This makes it possible to compare different projects, different development options for the same deposit, and different cash flow profiles within the same model.

From this logic comes NPV, net present value.

In practical terms, NPV shows what a project is "worth today" once future cash flows have been adjusted for time and discounted at a chosen rate.

At first glance, the idea seems almost unassailable. If future money is not equivalent to present money, then it does indeed need to be brought back to present value. In general, that is correct. In mining, this approach is especially useful because projects are capital-intensive, long-lived, highly sensitive to timing, and often do not begin returning capital until well after the initial investment period.

But this is exactly where one of the most underestimated problems in project evaluation begins.

Discounting is useful as a tool. But it is not neutral.

Everything depends on the discount rate.

And the discount rate is not a law of nature. It is an assumption, and therefore a judgement.

That is why NPV can be both a valuable analytical tool and a source of serious distortion if it is applied mechanically.

On paper, the procedure looks simple. Future cash flows are taken, discounted at the selected rate, summed, and an answer appears. In reality, however, the discount rate already contains an entire set of assumptions about risk, capital, time, project resilience, country conditions, technical complexity, market uncertainty, and the overall probability that the project will actually deliver the cash flow built into the model.

If those assumptions are poorly understood, chosen by habit, or inserted mechanically, the result may look tidy in form but be wrong in economic substance.

That is why the discount rate must never replace analysis of the project itself.

This is a fundamental point.

A well-built geological model can be undermined by poor economic assumptions.

In exactly the same way, a good project can be artificially weakened simply because someone inserted an inappropriate discount rate into the model without understanding what it implied.

In that sense, a poorly chosen discount rate can damage an evaluation no less seriously than a poor geological interpretation.

This is especially important in mining because we are not dealing merely with abstract future money. We are dealing with a mineral base that exists in physical space, develops through time, and in some cases may itself change in economic significance.

This is particularly visible in strategic commodities, supply-constrained commodities, and strongly cyclical commodities. In such periods, or in such market segments, mine-ready reserves, infrastructure-supported mining areas, high-quality material, and later project stages may retain, or even increase, their significance more strongly than a purely mechanical financial treatment would suggest. If very blunt discounting is applied to such projects, an important part of their economic value may be understated artificially.

That is why NPV should never be treated as a universal truth that automatically "tells the truth."

NPV is useful. But it does not remove the need to think.

If the discount rate is set too high, late cash flows lose weight in the model very quickly. As a result, long-life projects, deep projects, infrastructure-heavy projects, or deposits with a long but economically strong operating life may appear materially weaker than they really are. An excessively high rate can effectively cut away a large share of project value, especially where the later stages are not a weak tail, but remain industrially and commercially meaningful.

But the opposite extreme is just as dangerous.

If the discount rate is set too low, the project begins to look stronger and more attractive than reality permits. Late cash flows become too "heavy," risk is understated, capital recovery appears easier than it really is, and the project's vulnerability to time and uncertainty may be concealed.

In other words, the discount rate is not just a coefficient in a formula. It is one of the most powerful filters through which a project is viewed by investors, banks, mining companies, and analysts.

And if that filter is poorly chosen, the final conclusion may be wrong even if every intermediate calculation has been performed neatly.

That is why discounting must be treated seriously.

- Not as a mandatory formality.
- Not as a technical line in a spreadsheet.
- And not as a number that can simply be taken "from industry practice" without further thought.

The rate must be justified.

It should reflect:

- the stage of the project,
- the nature of the commodity,
- the expected life of the operation,

- country risk,
- the product market,
- technical complexity,
- capital structure,
- and the overall resilience of the cash flow.

Without that, NPV can very easily turn from an analytical tool into an instrument of self-deception.

At this point, it is worth returning once again to the main logic of the book.

Project economics is not one formula and not one number. It is a system of parameters. Discounting is only one of the ways of bringing the result of that system back to the present point in time.

- It does not replace analysis.
- It does not eliminate sensitivity.
- It does not repair a weak project.
- And it should never be used as a way of automatically "killing" a project that someone already dislikes intuitively.

The best way to treat NPV is as a useful but highly sensitive instrument. It helps compare projects and shows the value of time in cash flow. But it should be used only together with a clear understanding of what lies behind the discount rate, and whether that rate is appropriate for the nature of the project itself.

That is why, in sound mining evaluation, the question should never simply be: "What is the project's NPV?" as though that alone were enough.

The correct question is different:

What NPV is produced under which assumptions, at which discount rate, and how well justified is that rate for this particular project?

2.4 CAPITAL, MATURITY AND FINANCEABILITY

2.4.1 How Capital Looks at a Project

Up to this point, the project has been viewed mainly from the inside: first as a geological object, then as a source of potential product, and then as a system of cash flows sensitive to time, cost, price, discounting, and the overall robustness of its assumptions.

But at some point, every such line of reasoning runs into the next, very practical question: is anyone actually prepared to put money into this project?

This is the point at which the project is tested not only by geology, engineering, and economics, but by capital.

That is an important transition, because capital looks at a project differently.

A geologist may see a mineral system, its scale, continuity, and potential.

An engineer may see a mining method, a process route, and the limits of the technology.

An economist may see revenue, cost, sensitivity, and cash flow.

Capital, however, looks first at something else:

- risk,
- return,
- timing,
- resilience,
- predictability,
- and whether the project structure can be trusted at all.

That is why a technically strong object does not always become a project that is easy to finance.

Capital is interested not simply in the presence of a mineral commodity, and not simply in a positive NPV. It is interested in how real that result is, how well protected it is against errors in assumptions, how clearly the project logic is structured, and how high the risk is that money will be committed before the project begins to prove its viability.

That is why the phrase "the project has good economics" guarantees nothing by itself.

One must always ask: good economics for what kind of capital, at what stage of the project, under what risk profile, and over what investment horizon?

This matters especially because capital in mining is not homogeneous.

An owner, a strategic investor, a bank, an offtaker, a trader, a fund, an industrial consumer, or a state-backed player may look at the same object in completely different ways.

For one, the key issue may be growth in asset value and long-term upside.

For another, it may be security of repayment and protection of debt.

For a third, access to raw material and control over supply.

For a fourth, the opportunity to integrate the product into an existing processing or sales chain.

For a fifth, the political or industrial significance of the project, even if its pure commercial logic is less than perfect.

In other words, the same project may be highly attractive in one system of interests and of little value in another.

That is why capital never evaluates projects "in general." It evaluates them from the standpoint of its own role, its own objective, and its own tolerance for risk.

A bank, for example, is rarely interested in the geological beauty of the object as such. What matters to it is how reliable the cash flow is, how quickly the project begins to generate it, how robust the assumptions are, and what happens if the product price is lower than expected, costs are higher, or start-up is delayed. For a bank, downside protection, debt service capacity, and predictability of outcome are critical.

A strategic investor may look more broadly. It may be interested not only in current profitability, but also in long-term control over the mineral base, access to product, supply geography, the place of the project within its wider asset system, expansion potential, and the project's ability to strengthen the broader value chain.

A trader or an offtaker may look at the project primarily through product and marketability. It may care less about the "ideal economics of the model" and more about whether the project can reliably supply the required product, in the required volume, at the required quality, and into the required delivery point.

An early-stage equity investor may be willing to accept high risk if it sees substantial upside in project value. But the same object, once it reaches the stage of debt financing, will be judged by completely different criteria. At that point, potential matters less than control, reliability, and the quality of assumption verification.

That is why there is almost always a long distance between "the project looks promising" and "the project is financeable."

At early stages, capital may be willing to finance potential.

At more advanced stages, it demands confirmation.

As the project develops, the character of the question changes.

At first, the question is:

is there even an object worth paying attention to?

Then:

can it become an industrial project?

Then:

can it generate a durable cash flow?

And only after that:

is that cash flow reliable enough to support serious capital?

This is exactly where the quality of the entire project structure becomes critical.

Capital evaluates not only the numbers, but the degree of confidence that can be placed in them:

- How well founded are the price assumptions?
- How realistic are the cost assumptions?
- How well proven is the process route?
- How clear is the mining sequence?
- How resilient is the project to weaker market conditions?
- Does it have a real buffer?
- Does the team understand what the critical risks actually are?
- And perhaps most importantly: is the project assembled honestly as a system, or merely as a set of optimistic assumptions?

This is a very important point.

Capital can forgive complexity.

It can forgive a long development timeline.

It can even accept high risk.

What it tolerates very poorly is ambiguity, internal contradiction, and an artificially improved picture.

That is why a project evaluation prepared to produce a desired answer is almost always detectable. Formally, it may look neat. But if it lacks internal strength, capital begins to treat it not as a basis for decision, but as an additional source of risk.

This is one of the hardest truths of the mining industry: it is not enough for a project to be interesting. It must be convincing.

And in this context, convincing does not mean an attractive presentation or an impressive resource base. It means the ability to withstand uncomfortable questions:

- What happens if price is lower?
- What happens if costs rise?
- What happens if start-up is delayed?
- What happens if feed quality is less stable than expected?
- What happens if logistics fail?
- What happens if the later stages of the project are weaker than expected?

The better a project withstands those questions, the better it looks in the eyes of capital.

That is why, in the end, capital finances not only the object and not only the model. It finances the degree of trust in the project logic.

This is the point at which geological, technical, and economic work must come together as one integrated whole.

It is not enough to have a good object.

It is not enough to have a good product.

It is not enough even to have a good model.

The project must be assembled in a way that makes sense as a system:

- Where the money is created.
- Where value is lost.
- Which parameters are decisive.
- Which risks are critical.
- And by what logic the project is actually capable of returning invested capital.

Only then does the project begin to look not merely like a mineral base with potential, but like something capital is prepared to engage with seriously.

And that leads to the next step.

If capital views a project through risk, return, confirmation of assumptions, and the maturity of the overall structure, then the next thing to understand is how a project moves from an idea to the point at which it can be regarded as ready for a serious investment decision.

2.4.2 Project Maturity, Decision Points, and Capital Readiness

Once it becomes clear how capital looks at a project, the next natural question is this: at what point can a project actually be regarded as mature enough for a serious investment decision?

This is an important question, because a project does not become "ready" in a single moment. It passes through successive stages of maturity. At each of those stages, a different set of tasks is addressed, a different portion of uncertainty is removed, and a different level of confidence in the future project structure is established.

That is why a project should never be treated as something that either exists or does not exist in finished form. At the beginning, it is only a possibility. Then it becomes a hypothesis that still needs to be tested. Then a technical concept. Then an economic structure. Only after that does it begin to approach the state in which a serious capital decision can be made against it.

In that sense, project maturity is not defined by the name of the stage or by the volume of the report. It is defined by the extent to which the critical assumptions have been confirmed.

That is exactly why staging matters.

Not because corporate procedures happen to like it. Not because it is convenient to attach a formal label such as concept, pre-feasibility, or feasibility study. But because, at each stage, the project should close out its own key set of uncertainties. Until that has happened, it is too early to move to the next step.

At the earliest stage, the project exists only as a possibility. The main question here is simple: is there, in fact, an object worth further work? At this level, it is still too early to speak about detailed economics, but it is already necessary to understand whether the object has geological logic, whether it has a potential product, whether it could in principle interest the market, and whether the project idea is fundamentally compatible with basic technical and commercial constraints.

The next stage is conceptual development. Here the project stops being merely an interesting object and starts becoming a potential development concept. At this level, the key question is whether there is, in principle, a realistic path from the ground to a product. Not an ideal path, and not a final one, but a realistic one. Is it possible to envisage a mining method, a process chain, a basic logistics concept, and an overall commercial logic that are consistent with the nature of the object? If the answer to that question is no, there is little point in moving further.

After that, the project moves into the stage of comparative and preliminary evaluation of alternatives. Here the task is no longer simply to show that the object "can be done," but to understand which development option appears most viable. At this stage, possible schemes begin to be compared, parameters become more clearly defined, weak solutions are discarded, and it becomes clearer which project configuration is actually capable of passing both technical and economic scrutiny. This is where the project begins to take a shape that allows a more serious discussion about future capital.

Only after that can the project approach the stage of investment readiness. Here the main question becomes harder: have the key assumptions been confirmed to a sufficient degree for serious capital to be committed against them? Not "does the

TABLE 2.1
The Levels of Capital Investment Process

Project Stage	Main Question of the Stage	Indicative Time Horizon
Possibility	Is there, in fact, an object worth further work?	From a few weeks to 6–12 months
Conceptual development	Can a realistic industrial concept be assembled in principle?	Usually 3–12 months
Comparative and preliminary evaluation of alternatives	Which development option appears most viable?	Usually 6–18 months
Investment readiness	Have the critical assumptions been confirmed sufficiently for a serious capital decision?	Usually 12–24 months
Execution and start-up	Can the project deliver its promised value in practice?	From 1 to 2 years and upward, depending on scale

project appear attractive," but have the geology, mining sequence & processing route, product, logistics, costs, sensitivity, cash flow, and overall resilience of the structure been confirmed well enough that money would be committed not to a hope, but to a genuinely assembled system?

At this stage, one point becomes especially important.

Capital does not finance a report.

Capital finances the maturity of assumptions.

For illustration, the Table 2.1 provides an indicative view of project stages. It is not a rigid rule, but it shows the general logic of moving from possibility to investment readiness.

Actual timelines and stage content may, of course, vary materially depending on project scale, deposit complexity, level of knowledge, infrastructure, and access to capital. The table should therefore be read not as a formal standard, but as an order-of-magnitude guide.

If the geology is still too weak, the process route is not proven, the logistics chain is not assembled, the product market is described only in abstract terms, and costs exist only as average numbers, the project may look as attractive as one likes on paper, but that does not make it mature.

That is why every stage needs decision points.

They are needed not as a bureaucratic formality, but as protection against one of the most common errors in mining: pushing a project forward simply because time, money, and personal expectation have already been invested in it.

A decision point is the moment at which one must answer honestly: what has already been confirmed, what remains uncertain, which risks are still acceptable at the current stage, and whether it is possible to move further without self-deception.

At those points, the project must either:

- move forward,
- be reworked,
- change concept,
- or stop.

This is a very important principle.

A good project is not one that moves forward without interruption.

A good project is one that passes correctly through decisions.

Sometimes the most professional action is not to advance, but to stop. Sometimes it is to return to an earlier stage. Sometimes it is to abandon a convenient but weak concept in favour of a less obvious but more viable one.

That is why maturity stages cannot be skipped without consequences.

If a project behaves as though it were mature while in fact remaining immature, the consequences almost always accumulate later. Economics promised too early, project boundaries drawn too confidently, technological conclusions stated too firmly, a development concept fixed too soon, all these things create an early impression of progress, and later begin to work against the project.

Because what capital tolerates especially poorly is not complexity, but internal immaturity hidden behind apparent confidence.

Against that background, it is very useful to think about project maturity not as a linear ladder of reports, but as a sequence of questions that the project is obliged to answer.

First:

is there, in fact, an object worth attention?

Then:

can an industrial concept be assembled from it in principle?

Then:

which development option is actually better than the others?

Then:

how well have the critical assumptions been confirmed for serious capital to be committed against them?

And only after that:

is the project ready not merely to look good in a model, but to survive the transition into execution?

In that sense, maturity is not the volume of information. It is the quality of uncertainty removed.

A project may be described in great detail and still remain immature if the most important questions are still unresolved. And the reverse is equally true: a comparatively compact but honest and correctly assembled structure may be much closer to real investment readiness than a large model full of numbers but poor in confirmation.

This matters particularly because different stages call for different kinds of capital.

At an early stage, capital may be willing to finance potential.
At an intermediate stage, it begins to require demonstrable engineering logic.
At a later stage, it expects confirmed cash-flow resilience and manageable risk.

In other words, capital readiness is not a general state of "appealing" or "unappealing." It is the fit between project maturity and the expectations of the type of capital the project is trying to attract at that stage.

That is why the question should not be framed as: is the project ready for financing in general?

It should be framed as: is the project ready for the type of capital it is trying to attract now?

That is much more precise.

An object that is still too early for bank financing may be entirely interesting to a strategic investor. A project not yet ready for serious debt may still be attractive to risk capital. And an object that has already advanced too far to live only on a "potential story" may begin to require a much stricter level of confirmation.

This is the point at which a project ceases to be merely a geo-economic construct and becomes an object of discipline.

It is not enough to "feel" that the object is good.

It is not even enough to "believe" that it should work.

One must understand what stage it is at, which questions have already been resolved, which remain open, and on what basis it is actually possible to move further.

That is why project maturity is not a topic for a formal checklist alone. It is one of the main instruments of disciplined thinking in mining.

In the end, a project does not win because someone wrote the first beautiful story about it. It wins when, at the right moment, exactly the level of uncertainty required for the next step has been closed out.

And that leads to the next logical question.

If different stages require different types of capital and different levels of confirmation, then the next thing to understand is who finances those stages, why they do so, and why the same object may be attractive to one source of money and entirely uninteresting to another.

2.4.3 The Logic of Project Financing: Who Provides the Money, and for What

At some point, every mining project runs up not against geology, not against technology, and not even against the economic model, but against a much more practical question: who is actually going to provide the money?

At first glance, the answer seems simple. If the object is good, if it contains a mineral commodity, if the product is clear, and if the economics look positive, then capital should come. In real life, however, things are much harsher. Money does not come to a "good object" in the abstract. Money comes only when the project has been assembled into a structure that a specific source of capital considers acceptable for its own purposes.

That is why one of the most damaging illusions in mining is the idea that a project is somehow financed by its qualities alone. As if the existence of a resource, a product, and a positive model were already enough. In practice, that is almost never sufficient. Capital does not look at the project in our internal sense of the word. It looks at the project in its own.

A geologist sees the object.

An engineer sees the mining method.

A metallurgist or process specialist sees the treatment route.

An economist sees revenue, cost, and cash flow.

Capital sees something else: risk, controllability, return, timing, reliability of assumptions, and the probability of getting its money back with an acceptable outcome.

That is the central difference.

A project may be geologically strong, technically reasonable, and even economically attractive, yet still remain difficult to finance. Not because it is bad, but because its maturity, its risk structure, and the expectations of the specific capital source have not yet aligned.

This is especially important because capital in mining is never homogeneous.

Some money comes for growth in asset value.

Some comes for control over raw material.

Some comes for protected repayment.

Some comes for product and access to market.

Some comes for strategic influence, industrial integration, or broader political-economic interest.

That is why the question "is this a financeable project?" is too crude by itself. A more accurate question is different: for what kind of capital, at what stage, and under what risk profile can this project actually be financeable?

At the early stages, an object usually cannot promise reliable cash flow, because that cash flow does not yet exist. What it can promise is possibility: the possibility of discovery, the possibility of growth, the possibility of advancing to the next stage, the possibility of a future increase in asset value. That is why early money is almost always the most patient and, at the same time, the most risk-tolerant. It comes not for proven return, but for the chance to enter early and capture the largest increase in value if the hypothesis proves correct.

But the further the project advances, the less willing capital becomes to pay for story alone. As the object matures, money demands more confirmation and less belief.

At first, it is enough that the object deserves attention.

Then it must be shown that the industrial logic is not imaginary.

Then the development concept must be assembled and able to withstand scrutiny.

Then the cash flow must be not merely possible, but sufficiently dependable.

This is where an important shift takes place: the project stops being financed on potential and begins to be financed on proof.

In that sense, capital disciplines a project.
It forces the right difficult questions to be asked.
Not simply: is there metal, coal, or industrial mineral in the ground?
But:

- Can it be mined consistently?
- Can it be processed?
- Can it be sold?
- Can it be produced at the required scale?
- Can it be delivered on the required timeline?
- Aan it withstand deterioration in conditions?
- And, ultimately, can the invested money be returned?

This is where it becomes clear why the same object may be highly attractive to a strategic partner and of little interest to a bank.

A strategic investor may look far ahead. What matters to it may be access to raw material, the project's place in the supply chain, strengthening of its industrial base, or control over future product. It may tolerate a longer horizon and a higher level of uncertainty if it sees structural value in the object.

A bank looks differently. It rarely wants to share a geological dream. What matters to it is whether the project can service obligations, how reliable the assumptions are, how well protected the cash flows are, what happens if the market weakens, and how quickly model error can become a repayment problem.

A trader or product buyer may look at the project differently again. It may care less about an "ideal NPV" than about whether the project can reliably deliver the required product in the required volume, at the required quality, and into the required delivery point. For such a party, the value of the project may lie less in formal profitability than in control over physical material flow.

That is why capital never finances an object "in general." It finances the version of the project that it understands and accepts within its own logic.

A hard but liberating conclusion follows from this: it is not enough for a project to be good; it must be legible to the right type of capital.

Not legible in general.

Not merely "attractive overall."

But legible to the specific source of money the project is actually trying to attract at that stage.

At this point, it is often more useful to have an internal checklist than a beautiful story. Not a banking list of a hundred items, and not corporate bureaucracy, but a short set of questions that quickly shows whether the project is genuinely ready for a serious conversation about money.

2.4.3.1 Checklist: Is the Project Ready for a Capital Conversation?

What exactly is being financed? An idea, confirmation of a hypothesis, transition to the next stage, construction, start-up, expansion, stabilization of operations? Until that is stated precisely, the financing discussion remains vague.

What is the project's main risk at this stage? Geology, processing, product, market, logistics, construction cost, operating resilience, country exposure, team quality? If the main risk cannot be named, it is usually the real risk.

Which assumption is still the weakest? Not the most visible one, but the weakest one, the place where the model still depends more on belief than on confirmation.

What has already been confirmed, and what is still only assumed? Those two categories must not be mixed. Many weak projects fail precisely where assumptions were presented as facts.

Do we understand who will buy the product, and why from us? Not "there is a market," but who exactly, on what terms, against which specification, and against what alternatives.

Do we understand where the project loses money first? In mining, processing, logistics, commercial terms, product quality, or capital structure? When a project destroys value, it usually does not do so everywhere at once. It starts in one weak link.

Does the project have a resilience buffer, or does it live on a narrow set of optimistic assumptions? A good project can withstand uncomfortable questions. A weak one starts to crack at the first movement in price, cost, or schedule.

Does the current stage of the project match the type of capital being sought? This is one of the most underestimated questions. Very often a project fails to raise money not because it is bad, but because it is asking the wrong capital at the wrong moment.

Do we understand why the investor or lender would want this project at all? Asset growth, product access, control over raw material, debt repayment, industrial synergy, political significance? Until the financing party's motive is understood, the conversation remains one-sided.

Can the project withstand an uncomfortable conversation? This is almost the main question. If, after a series of hard questions, the logic of the project starts to fall apart, then it is not yet ready for serious capital, even if the model looks tidy on the surface.

This checklist is useful because it quickly pulls the discussion back from the level of polished presentations to the real state of the project. In practice, money rarely goes where there is only enthusiasm. It goes where uncertainty has either already been reduced, or at least honestly identified and correctly allocated.

That is why project financing is not a final act, not a reward for good work, and not a formal continuation of techno-economic evaluation. It is a separate stage of maturity testing.

At that point, the project is not being asked how interesting it is in theory. It is being asked how well assembled it is, how well it can withstand doubt, and how clearly it is understood what must happen for the invested money to come back.

In the end, capital finances neither the resource, nor the report, nor the presentation. It finances the degree of trust in the project structure.

That is why the main question should never be framed simply as:

where do we find the money?

The prior question is different:

what type of money makes sense for this stage, and is the project mature enough that those funds will not be frightened off before we have even had the chance to convince them?

And that leads directly to the next step. If different types of money enter the project for different reasons and at different stages, then the next thing to understand is how project finance structure is built, which money comes first, which comes later, and why the cost of capital changes as the object itself becomes more mature.

2.4.4 Project Financing Structure

Once it becomes clear that different stages of a project correspond to different types of capital, the next practical question is this: how is project financing structure actually built over time?

This matters because a mining project almost never passes through its full life on the same type of money. Different stages attract different capital sources, and each of those sources enters for its own result, on its own terms, and at its own price for risk.

That is why project financing structure is never a random collection of money. It reflects the maturity of the object, the level of uncertainty, and the extent to which the project has already demonstrated its viability.

At the earliest stage, money is almost always the most expensive.

Not because anyone is trying to "punish" the project for being immature, but because risk is highest at that stage. The geology may still be only partly confirmed, the development concept may not yet be fixed, the process route may not be assembled, the product market may still be insufficiently defined, costs may remain largely unproven, and it may still be unclear whether the object can successfully pass into the next stage.

That is why early capital finances uncertainty above all else.

And that is exactly why it demands its price.

For that type of capital, what matters is not a stable cash flow, but the possibility of a significant increase in value if the project succeeds. In other words, early money enters not because the project has already become a durable business, but because it still contains room for rapid re-rating.

- As the project develops, the picture begins to change.
- Some uncertainties are removed.
- The geology becomes clearer.
- The process route becomes more defined.
- The product and its market begin to take more concrete shape.
- Costs become less abstract.
- The development concept stops being just an idea and becomes a more assembled structure.
- And with that, the perception of risk changes.

This is the point at which other types of capital may appear. More cautious capital, less tolerant of raw uncertainty, but also less expensive. Such capital no longer comes for a value-growth story at any price. It comes when it can see that the project is moving from hypothesis toward a more confirmed system.

This is one of the central regularities of mining project finance: the greater the maturity of the project, the wider the circle of accessible capital and, as a rule, the lower its cost.

But that rule works only when maturity has actually been achieved, not merely declared.

If a project tries to move too early from expensive money to cheaper money without first closing its key uncertainties, it almost always encounters one of two outcomes. Either the capital does not come at all. Or it comes on terms that make it clear the market still sees the risk as high.

That is why financing structure should be viewed not as a formality, but as a reflection of the project's real condition.

At the earliest stages, a project is usually funded by owners, risk capital, early-stage investors, or strategic partners prepared to accept high uncertainty. That is logical. Such capital understands that its role is to finance the path from possibility to a more mature stage.

At intermediate stages, strategic investors, industry partners, future product buyers, traders, or other participants may begin to appear, for whom the value of the project lies not only in financial return as such, but also in access to raw material, product, or future supply flow.

At later stages, once the project has been assembled into a sufficiently mature structure, debt capital may become available. But the word *may* matters here. Debt does not come simply because the object is good. It comes only where it is already reasonably clear how that debt will be repaid.

That is why debt financing almost always imposes the strictest requirements on the quality of assumption confirmation.

For debt, it is no longer enough to have:

- interesting geology,
- a general industrial logic,
- a potentially good market,
- or even a model that appears positive.

Debt requires:

- reliable cash flow,
- understandable contracts,
- manageable risks,
- a predictable cost structure,
- and a sufficiently high level of confidence that the project can survive weaker conditions.

In other words, each successive type of capital requires not simply a "better project," but a more mature project.

This leads to another important conclusion: the cost of capital does not change by itself. It changes together with the project.

While the object is still raw, money is expensive.

As the project becomes clearer and more robust, money usually becomes cheaper.

When provable cash flow begins to emerge, some forms of capital become more accessible still.

This does not mean that a mature project always receives cheap money. But it does mean that, as uncertainty is reduced, the project usually gains access to a broader and more demanding, yet less expensive, pool of capital.

That is why one of the key tasks in project development is not only to attract financing at all, but to improve the quality of that financing over time.

A strong project is not simply an object that once managed to find money. It is an object that, as it matures, is able to move from the most expensive and risk-tolerant capital toward a more stable and less expensive financing structure.

This is a very important practical point.

If a project remains dependent on the most expensive money for too long, that almost always means one of two things. Either it is not maturing fast enough. Or risks remain in the structure that do not allow the cost of capital to fall.

In both cases, that is a signal.

A good object should not merely survive on capital. It should gradually improve its ability to speak to more demanding, but less expensive, sources of money.

Of course, in real life financing structures are often mixed.

A project may contain, at the same time:

- owner's equity,
- a strategic partner,
- risk capital,
- preliminary commercial agreements,
- debt instruments,
- support from a product buyer,
- state-backed mechanisms,
- or various hybrid forms of participation.

But even in such mixed structures, the logic remains the same.

Each source of money is covering not some abstract "project need," but a particular stage of risk, and each expects a particular type of result.

That is why financing structure must be aligned with project maturity. Otherwise, distortion appears.

Capital that is too expensive at too late a stage starts to weigh on the project and erode its economics.

Money that appears "cheap," but is brought in too early, is usually either unavailable or attached to conditions that quickly reveal that risk was underestimated.

The wrong balance between capital sources can make even a strong object difficult to develop.

A deeper conclusion follows from this.

Project finance is not merely a question of whether money exists. It is a question of sequence.

- Which money should come first?
- Which risks should it be prepared to take?
- What exactly must the project prove in order to move to the next type of capital?
- And does the project remain too long in the zone of expensive money simply because it has failed to close its key uncertainties in time?

This is where real project maturity reveals itself. Not in the amount of money raised as such. But in the project's ability, step by step, to improve the quality of its capital. Because in the end, a strong project is not simply one that someone agreed to fund. A strong project is one that can reduce the price of money as it becomes more mature, more understandable, and more robust. This is the point at which financing structure stops being a secondary financial wrapper. It becomes part of the project's economics itself. And that leads directly to the next logical question.

If money at different stages comes at different prices, with different expectations and different requirements, then the next thing to understand is what mistakes are most often made when a project is packaged for capital incorrectly, too early, or too optimistically.

2.4.5 From a Good Project to a Financeable Project

A good project and a financeable project are not the same thing.

At first glance, that may sound unnecessarily severe. If the object is geologically convincing, if a reasonable process route has been assembled, if the product is clear, the market exists, and the economic model shows a positive result, it seems natural to expect that capital should come. In real life, however, there is almost always a distance between a quality project and a project capable of attracting serious money.

That distance is exactly what defines how mature the project really is.

A good project may have a strong mineral base. It may show reasonable economics. It may be attractive in terms of product, market, or long-term potential. But it becomes financeable only when the key elements of its structure begin to be seen as sufficiently confirmed, interconnected, and manageable.

In other words, capital does not finance a positive story as such. It finances a project that can withstand a test of coherence.

This is a very important point.

In mining, weakness rarely lies in one single fatal flaw. Much more often, the problem is different: taken individually, many elements appear acceptable, but together they still do not form a structure strong enough to inspire real confidence.

The geology may be interesting, but the process route is still unproven.

The process route may look reasonable, but logistics remain weak.

The product may be understandable, but the market is described too broadly.

The economics may be positive, but the project may be too sensitive to price or cost.

The stage of maturity may call for one type of capital, while the project is already trying to speak to another.

In all such cases, the object may be good in essence, but not yet financeable in the relevant sense.

That is why the path from a good project to a financeable project is not a matter of cosmetic improvement of the model. It is a matter of systematically removing uncertainty.

- First, the object itself must be understood.
- Then, how that object becomes a product.

- Then, how that product becomes revenue and cash flow.
- Then, which parameters in the system are actually decisive.
- Then, how those parameters change over time.
- Then, how capital looks at the project.
- Only after that does it become possible to speak meaningfully about financeability.

In that sense, financeability does not arise from the desire to attract money. It arises from the maturity of the project structure.

This matters for another reason as well. Capital almost never punishes a project simply for being complex. Complex objects get financed. Long projects get financed. Technically unconventional solutions may also get financed. What capital tolerates very poorly is internal ambiguity, weak coherence, and any attempt to replace confirmation with enthusiasm.

This is the point where the language of presentations ends and the language of trust begins.

A financeable project is one in which it is clear:

- What is being mined.
- How it is being mined.
- What it is being turned into.
- To whom it is being sold.
- Where money is lost.
- What holds profitability together.
- Which risks are critical.
- And why the whole structure should survive the transition from model to reality.

Until those questions have clear answers, it is premature to speak about financeability, even if the project looks attractive on the surface.

Another important point follows from this.

A financeable project is not necessarily an "ideal project." It is a project that has reached a sufficient level of maturity for the type of capital it is trying to attract.

For early-stage capital, that may mean one thing.

For a strategic investor, another.

For debt financing, something else again.

Financeability is therefore always relative to stage, risk structure, and source of funds.

That is why one cannot speak meaningfully about financeability without also asking: who exactly is expected to provide the money, and for what exactly are they prepared to provide it?

If that question is not asked, the discussion remains too abstract.

If it is asked correctly, the project begins to be seen not as a set of attractive characteristics, but as a system that can be matched to a specific type of capital.

This is one of the clearest signs of mature project thinking.

It is not enough to know that the object is good.

One must understand where it is still immature.

It is not enough to see strong economics.

One must understand how well those economics are actually confirmed.

It is not enough to calculate a positive result.

One must understand under which assumptions that result remains valid.

It is not enough to assemble a convincing story.

One must assemble a structure capable of withstanding someone else's doubt.

Because in the end, the project that becomes financeable is not the one that looks most impressive. It is the one that withstands a serious conversation better than the others.

That is why a good project does not become financeable at the moment a beautiful report is written about it. It becomes financeable when there are no longer critical internal gaps between its geology, process route, product, market, economics, risks, and stage of maturity.

That is the real boundary between a merely good object and a project capable of having a serious conversation with capital.

And if one looks back at the whole path covered so far, it becomes clear that financeability does not appear at the end as a random bonus. It is the result of all the work that came before it.

First, the object had to become understood.

Then described.

Then quantified.

Then translated into product and cash flow.

Then tested against sensitivity, time, mining sequence, and resilience.

Only after that does it begin to take a form in which capital sees not only potential, but a logic of risk and return that is acceptable to it.

In that sense, a financeable project is not a new entity, separate from geology and economics. It is a project in which geology, engineering, market logic, and capital begin, for the first time, to speak the same language.

3 Practical Application and Decision Tools

3.1 RAPID PROJECT READING

One of the most common mistakes at first contact with a mining project is to let attention jump immediately to tonnage, average grade, attractive drill intercepts, or the total size of the resource. All of that can look impressive, but none of it, by itself, answers the central question: what is the economic essence of this object, and where is the real boundary between its strength and its weakness?

That is why a rapid first-pass evaluation should not try to understand everything at once. What matters far more is to identify, in a short time, a few essential things: what in this project could actually create value, what could destroy that value, and which parameter is most likely to be the critical constraint.

That is the real purpose of a quick project read. It is not meant to replace full analysis. It is meant to set priorities correctly. In other words, the first task is to understand where to look first, what is critical in this project, what is still secondary, and what can temporarily be left aside at the early stage without losing the core meaning of the object.

The first question should not be about tonnage or average grade. It should be a harder one: what exactly is supposed to become the saleable product here, and how realistic is that?

For one project, the answer may be an iron ore concentrate of a given specification. For another, a copper concentrate. For a third, gold recovered through a particular process route. In a coal project, coal or a prepared product meeting the buyer's specification. In an industrial minerals project, a product defined by chemistry, whiteness, particle size, or other performance characteristics.

Until that is clear, discussion of the project is usually premature. The market does not pay for the resource as such and does not pay for the geological model. It pays only for a product brought to the required specification. So, on first reading, one should quickly understand: what product is being proposed, how achievable it is, and whether a simple geological story is hiding a difficult and expensive task of converting raw material into a marketable product.

The second question should follow immediately: what matters most here, mining, processing, market, or logistics?

The main risk sits in different places in different projects.

In one case, the geology is simple, but the product is difficult to sell.
In another, the market exists, but processing is complex and unstable.
In a third, processing is clear, but logistics destroys the economics.
In a fourth, everything looks acceptable until it becomes clear that the project is overly sensitive to one impurity, one ore type, or one price assumption.

 DOI: 10.1201/9781042018932-3

That is why a good rapid review does not begin by checking everything. It begins by trying to identify the project's main knot. In most cases, that knot determines the project's fate.

A simple rule is useful here. Almost every mining project has one question that matters more than all the others. Until that question is clarified, detailed work on secondary parameters changes very little.

In an iron ore project, that key question is often not tonnage itself, but ore type, processing behaviour, and the quality of the future concentrate. High grades alone do not guarantee strong economics. In iron ore, phosphorus and sulphur quickly become critical because they can sharply reduce product quality and restrict the buyer universe. It is equally important to understand what kind of ore this is: magnetite, hematite, or mixed. Magnetite often supports a clearer magnetic beneficiation route. Hematite ore frequently requires a more complex and less stable upgrading route. So, at first pass, what matters more than total resource is the combination of feed quality, deleterious impurities, expected concentrate yield, achievable product specification, and logistics to market.

In a copper project, the first issue is usually the type of mineralization and the degree of ore consistency. If the project is based on fairly uniform sulphide ores that float well, it may look relatively straightforward. But if mixed ores, transitional material, oxide zones, or a variable sulphide-oxide balance are present, the picture changes completely. Formally, the average copper grade may still look acceptable, but treatment becomes more difficult, recovery falls, the process route may split into different streams, and a large share of value can be lost in the transition from grade to product. In a rapid copper review, it is therefore often more important to ask not simply what the average copper grade is, but what kind of ore this is and how consistently it will behave in flotation.

In a gold project, the central issue is often ore refractoriness. This is one of the most dangerous cases in which attractive grades create the illusion of a strong project, while the mode of occurrence of the gold makes recovery much harder. If carbonaceous material, preg-robbing behaviour, or sulphides such as arsenopyrite or pyrite with finely locked gold are present, the real economics may depend less on grade than on the need for liberation and specialized pre-treatment. That may mean pressure oxidation, roasting, bio-oxidation, or some other complex stage that radically changes CAPEX, OPEX, technical risk, and scale requirements. So, in gold, the fast question is not "how many grams per tonne?" but "how accessible is this gold technically, and at what cost?"

In a coal project, the first limiting factor is often not the coal itself, but distance to market and logistics. A deposit may have acceptable quality characteristics, but if it is remote from consumers and lacks access to rail, port, or inland water transport, logistics can destroy the economics before the detailed analysis even begins. In coal, it is dangerous to look only at ash or calorific value. Logistics, consistency of quality, saleable yield after preparation, and proximity to the customer are just as important. In that sense, the essence of a coal project is often understood faster from a map than from an assay table.

The same logic applies to other commodities. In industrial minerals, the key issue may be not total volume, but whiteness, chemical purity, reactivity, particle size, deleterious impurities, or the narrowness of the end market. In construction materials, what matters is often not the "valuable component" at all, but haul distance, proximity to market, and delivered cost.

A very practical conclusion follows from this. To grasp the essence of a project quickly is not to memorize more numbers. It is to identify as early as possible where the main value filter sits.

Sometimes that filter is processing.
Sometimes logistics.
Sometimes the product itself.
Sometimes one impurity.
Sometimes ore type.
Sometimes mining sequence.
Sometimes the cost structure.

And sometimes the market, which appears to exist in theory but is too narrow or too demanding in specification in reality.

After that comes the next layer of quick evaluation: what should be looked at first, and what can be set aside temporarily?

In most cases, the first things to examine are these:

First, what product is being proposed, and can it actually be made?
Second, which properties of the ore or raw material determine processing difficulty?
Third, which impurities, ore types, or technological features are capable of destroying the economics?
Fourth, is there a real market for this exact product, not for some abstract analogue?
Fifth, what role does logistics play, and does it consume the project?
Sixth, which parameter is most sensitive: price, recovery, concentrate yield, impurities, transport, mining sequence, or something else?

Only after that does it make sense to go deeper into the classic indicators: tonnage, average grade, detailed resource classification, elaborate NPV calculations, and second-order sensitivities.

That does not mean those things are unimportant. They are important. But at an early stage of acquaintance with a project, they are often secondary to the main constraint. Put more simply, there is little value in spending a long time analyzing a beautiful resource base if it is already obvious that the product is problematic, processing is unstable, or logistics is fundamentally difficult.

The reverse side is also important. What can temporarily be ignored at an early stage?

If the task is to grasp the essence quickly, rather than prepare a feasibility study, then one does not initially need excessive detail in the block model, fine variography, second-order discounting scenarios, overly elaborate financing structures, or formally tidy refinements that still do not resolve the main issue. All of that has its place later, once the main product, the main risk, and the main value filter are already understood.

In other words, at the early stage it is often reasonable to postpone whatever does not change the central conclusion.

If, in an iron ore project, it becomes clear in the first hours that phosphorus, sulphur, expected concentrate yield, and logistics are decisive, then secondary debates about resource classification are not yet the priority.

If, in a copper project, the key problem is mixed ore and unstable flotation, there is no point pretending that average grade already says enough.

If, in a gold project, the central uncertainty is refractoriness and the need for expensive pre-treatment, then that is what must sit at the centre of attention, not some abstract "average project economics."

If, in a coal project, logistics is critical, then that is what should be tested before anyone starts admiring reserve volumes.

That is why one more rule is useful in rapid project review: do not look first for everything. Look first for the project's main contradiction.

That contradiction usually reveals its true essence.

Sometimes it is the contradiction between good geology and a weak product.
Sometimes between high grade and poor processing behaviour.
Sometimes between a clear product and expensive logistics.
Sometimes between strong early economics and weak long-term resilience.
Sometimes between an attractive model and overly optimistic assumptions.

That contradiction is precisely the place where a project either begins to turn into a business, or remains only a good geological story.

The practical meaning of a quick project read can therefore be reduced to a few working qucstions:

What is actually being sold here?
What must happen for the material in the ground to become a saleable product?
What single factor is most capable of destroying the economics?
Which parameter matters more than the others in this case?
And what can still be left temporarily uncomplicated at this stage without losing the main point?

If those questions are asked correctly, the essence of the project usually becomes clear much earlier than the end of the presentation or technical report.

3.1.1 Brief Practical Conclusion

To understand the essence of a mining project quickly does not mean memorising as much data as possible. It means identifying as early as possible:

- What product is supposed to be produced.
- What most strongly determines whether that product is achievable.
- Where the main risk lies.
- Which parameter is critical.
- And what, in this object, is most likely to destroy value first.

That is where professional evaluation begins. Not with admiration for the numbers, but with an understanding of where the project's main economic knot actually sits.

3.2 STRUCTURED EVALUATION CHECKLISTS

The checklists set out below are intended for practical evaluation of mining projects. Their purpose is not to replace detailed technical, economic, or investment analysis, but to help structure a first reading of a project quickly, identify the main questions, and avoid missing critical weak points.

The checklists are arranged by level. Some help establish quickly whether the project has basic economic logic at all. Others are designed to test the geological foundation, raw material quality, process route, market, logistics, cost structure, sensitivity, project maturity, and readiness for capital.

They should not be used mechanically, but as working tools. The point is not the number of boxes ticked, but the ability to see as early as possible where the project is genuinely strong, where it is vulnerable, and which questions require immediate attention.

3.2.1 Rapid Checklist: Does the Project Have Basic Economic Logic?

Table 3.1 presents a rapid checklist intended for the first-pass review of a project.. Its purpose is to understand quickly whether the object has a basic economic logic, or whether one of the project's key supports is already missing at the outset.

3.2.1.1 Practical Conclusion

If, after the first review, it is clear what will be sold, how it will be produced, who needs it, and where the cash flow is supposed to come from, the project is worth taking further. If one of those elements remains unclear, the priority is to clarify the project's foundations, not to deepen the calculations.

3.2.2 Geological Checklist

This checklist is intended to show quickly whether the geological foundation of the project genuinely supports its economics, or whether the whole story rests on an interpretation that is too weak or too optimistic.

3.2.2.1 Practical Conclusion

The purpose here is not to re-check the entire geology immediately, but to understand whether the project stands on a real geological foundation. If the model is coherent, uncertainties are named honestly, and the data base is appropriate to the stage, geology supports further analysis. If the project looks more precise and mature than its actual level of knowledge allows, that is one of the first signs of a weak structure.

3.2.3 Checklist on Raw Material Quality and Ore Types

This checklist is intended to show quickly how homogeneous the material is, which of its properties determine processing and product quality, and whether an apparently acceptable average grade is concealing a more complex and less stable reality.

TABLE 3.1
Rapid Project Checklist

Check Point	What to Look At	Warning Sign
What product is being proposed?	Concentrate, Direct Shipping Ore(DSO), metal, coal product, industrial mineral of defined specification	The product is not named, or is described too vaguely
Is there a clear path from the ground to the product?	Mining, processing, preparation, quality requirements	There is a resource, but no clear route to a saleable product
Is there a market for this exact product?	Who the buyer is, what specification is required, how price is formed	The project speaks about "a market," but does not explain who will pay, and for what
Is there one obvious factor that could destroy the project?	Impurities, difficult processing, logistics, water, power, narrow market, regulatory constraints	A bottleneck is visible from the start, but the project avoids discussing it
Is it clear where cash flow is supposed to come from?	How grade turns into product, revenue, and margin	The discussion focuses on resource and quality, but not on the mechanics of making money
Does the story look better than its physical and commercial basis?	The relationship between presentation, actual data, technology, and market	Strong presentation, weak internal logic
Is the project scale realistic?	Mining scale, processing scale, logistics scale, capital scale	The economics only work at an overly ambitious or weakly justified scale
Does the project depend on an overly optimistic assumption?	Price, recovery, concentrate yield, transport, CAPEX, OPEX	One parameter is clearly stretched, and without it the project weakens
Is there an internal logic to the project?	The relationship between geology, processing, product, market, and economics	Individual elements exist, but do not form one coherent system
Does the object deserve deeper analysis?	Overall impression after checking the key supports	At least one basic element of the project is already missing at first pass

TABLE 3.2
Geological Checklist

Check Point	What to Look At	Warning Sign
What type of deposit is this?	Stratiform, vein, porphyry, skarn, BIF, placer, etc.	The deposit type is not clearly defined or is described too superficially
How geologically predictable is the mineralization?	Expected continuity, variability, body morphology	Bold conclusions are being drawn for a complex object from a weak data base
Is there a coherent geological model?	Structural logic, position of ore bodies, controlling factors	There is a dataset, but no coherent model

(Continued)

TABLAE 3.2 *(Continued)*
Geological Checklist

Check Point	What to Look At	Warning Sign
How homogeneous is the mineralization?	Variability in grade, thickness, structure, ore types	Average values conceal strong internal heterogeneity
Where is the main source of geological uncertainty?	Continuity, structure, domaining, faults, variability	Uncertainty exists, but is barely discussed in the project
Is the data density sufficient for the current stage?	Drill spacing, sample distribution, sample quality	The project is presented as mature, although the factual basis is still weak
How robust is the domaining?	Geological, structural, ore, and processing domains	Domains are either too coarse or artificially over-detailed
Does the model create a false sense of precision?	Complex 3D models, detailed blocks, precise numbers on limited data	The model looks convincing visually, but the data do not justify that level of confidence
Is the project economics critically dependent on poorly understood zones?	Best grades, main tonnage, key project areas	Most of the project's value rests on a weakly confirmed part of the object
Does the transition from geology to resource estimate look sound?	Interpolation logic, interpretation boundaries, reasonableness of extrapolation	The resource looks "too good" relative to the quality of the underlying data

TABLE 3.3
Material Quality Checklist

Check Point	What to Look At	Warning Sign
What ore types or raw material types are present?	Magnetite, hematite, mixed iron ores; sulphide, oxide, mixed copper ores; free-milling and refractory gold; different coal grades; different quality classes of industrial minerals	The project speaks about "one ore," while several fundamentally different material types are actually present
How homogeneous is the material?	Variability in composition, mineralogy, impurities, physical properties	Average figures look good, but hide strong internal heterogeneity
Are there transitional or problematic zones?	Mixed ores, transitional types, weathered zones, refractory zones, unstable quality intervals	Such zones exist, but their impact on processing and economics is barely addressed
Which impurities are critical?	For iron ore: P, S, Al_2O_3, SiO2; for coal: sulphur, ash, moisture; for gold and base metals: As, Sb, organic carbon, etc.	Impurities are present, but the project focuses only on the main component
How does ore type affect processing?	Magnetite versus hematite, sulphides versus oxides, free versus finely locked gold, easily beneficiated versus difficult coals	Ore type is mentioned, but its practical significance for processing is not explained
How variable is material quality over the mining sequence?	Distribution of ore types and impurities by depth, flanks, mining phases	Good zones are placed early, while more difficult material is pushed later without discussion of consequences

(Continued)

TABLE 3.3 *(Continued)*
Material Quality Checklist

Check Point	What to Look At	Warning Sign
Can the material be blended effectively?	Blending flexibility, plant feed stability, quality variability	The project requires excessive selectivity or depends on unstable feed
Is there a risk that the same average grade hides different economic value?	Differences in mineralogy, recovery, product yield, impurities	Ores with the same grade actually generate different economic outcomes
Is the project dependent on the "best" ore type?	Share of magnetite versus hematite, sulphides versus mixed ores, free gold versus refractory material	The economics look strong only while the most convenient material dominates the model
Is the link between ore type and processing clear?	How mineralogy, impurities, oxidation state, refractoriness, and physical properties affect flowsheet, recovery, and product quality	Geology is described separately, technology separately, but their practical link is not explained

3.2.3.1 Practical Conclusion

At this stage, the key is not simply "what is the grade," but what material will actually enter mining and processing. For iron ore, ore type and impurities must be checked immediately. For copper, the homogeneity of sulphide ore and the presence of mixed zones. For gold, refractoriness and the mode of gold occurrence. For coal, quality, saleable yield, and consistency along the logistics chain. If material types differ fundamentally, they cannot be hidden behind one average number.

3.2.4 Processing Checklist

This checklist is intended to show quickly whether there is a realistic process route for the object, how well it is supported, and whether acceptable laboratory results are concealing a more complex and expensive production reality.

3.2.4.1 Practical Conclusion

The key question is not simply whether something can be extracted from the ore, but whether there is a realistic, confirmed, and economically viable processing route. If the technology is clear, supported by representative testwork, and robust to material variability, that is a strong support for the project. If the route depends on limited tests, averaged recoveries, or overly "ideal" assumptions, processing risk may be one of the project's main constraints.

3.2.5 Product and Market Checklist

This checklist is intended to show quickly whether the project has not just a potential product, but a real saleable product with a clear market, clear quality requirements, and an achievable realized price.

TABLE 3.4
Processing Checklist

Check Point	What to Look At	Warning Sign
Is there a clear process route?	Crushing, grinding, beneficiation, flotation, gravity, hydrometallurgy, pyrometallurgy, autoclave, bio-oxidation, etc.	The process route is described in general words, but without a clear working flowsheet
Is the route appropriate for this exact material?	Fit between flowsheet and ore type, mineralogy, impurities, refractoriness, size distribution, moisture	A "standard flowsheet" is used, but there is no explanation why it should work here
Is the technology standard or non-standard?	Simple, proven route versus complex multi-stage treatment with elevated technical risk	The project relies on a complex route but presents it as almost routine
Are the key parameters supported by testwork?	Laboratory tests, pilot tests, variability testing, representative samples	Major conclusions are based on limited or non-representative testing
Is the testwork representative?	Coverage across ore types, impurities, zones, and material variability	Good results were achieved only on "easy" samples
What are the real recoveries?	Metallurgical recovery, concentrate yield, stage losses	The project quotes only "best" recoveries without range or context
How dependent is recovery on ore type?	Differences across sulphide, oxide, mixed, refractory, and other material types	The model uses one average recovery even though the material is clearly heterogeneous
Is there a critical dependence on fine grinding, selectivity, or reagents?	Liberation size, reagent consumption, flowsheet sensitivity	Acceptable results are obtained only under overly tight or expensive conditions
Are different processing streams required?	Separate treatment of ore types, pre-separation, different operating regimes	The object contains several material types, but the flowsheet is presented as single and simple
Is there a risk that the route is technically possible but economically weak?	CAPEX, OPEX, energy intensity, operating complexity, scale requirements	"It works in the lab," but the cost of the route is barely discussed
How robust is the route to feed variability?	Variation in feed, impurities, moisture, particle size, ore composition	The process works only on stable and "easy" material
Is it clear what product the route actually delivers?	Concentrate quality, impurity levels, moisture, yield, commercial suitability	The route appears to deliver a product, but its marketable specification is not disclosed

3.2.5.1 Practical Conclusion

The key question is not whether there is a price for a similar commodity, but whether the project has a real saleable product with a clear specification, a buyer, and an achievable realized price. If the product is commercially clear, the market is understandable, and the pricing logic is consistent with quality and logistics, that is a strong support. If the market is described abstractly, the price is arbitrary, and product requirements are vague, the commercial side of the project remains weak.

TABLE 3.5
Product and Market Checklist

Check Point	What to Look At	Warning Sign
What exactly will be sold?	Concentrate, DSO, metal, coal product, industrial mineral, chemical feedstock of defined specification	The product is described too vaguely or not defined at all
Is the target product specification clear?	Main component, impurities, moisture, sizing, whiteness, reactivity, ash, calorific value, etc.	The project speaks about a "quality product," but without clear parameters
Who is the potential buyer?	Smelters, traders, power stations, chemical plants, construction market, specialist consumers	The project talks about "the market," but does not name a real buyer type
Is there a market for this exact product?	Not for an abstract analogue, but for the specification that is realistically achievable	The project assumes a market for a better product than it can actually make
How is price formed?	Exchange-based, index-based, contract, negotiated, regional	The model uses a "market price" without explaining how it is formed
Are there material commercial deductions?	Penalties, impurity discounts, moisture deductions, sizing adjustments, transport and trading deductions	The model uses a gross headline price without proper commercial deductions
How sensitive is the product to impurities?	P and S in iron ore, As and Sb in concentrates, ash and sulphur in coal, impurities in industrial minerals	The main component looks good, but the impact of impurities is barely addressed
How important is quality consistency?	Permissible variability, blending capability, buyer requirements	The project assumes an unstable product, but this is not reflected in the commercial logic
Is the market broad or narrow?	Number of potential buyers, substitutability, market liquidity	Sales depend on one or two counterparties or on a very narrow segment
Is there a risk of confusing price on paper with price in a real deal?	Netback, real FOB/CIF (Free on Board (FOB), Cost, Insurance and Freight (CIF))price, quality, parcel size, remoteness	A "beautiful" reference price is used without linking it to real sales conditions
Does the product require additional preparation before sale?	Drying, pelletizing, sizing, quality stabilization, blending, packaging	The model assumes a ready product, but required finishing steps are not included
Is it clear why this product should be competitive?	Quality, price, logistics, availability, niche application	The product exists, but it is unclear why the market should choose it

3.2.6 Logistics Checklist

This checklist is intended to show quickly whether logistics is destroying the project. For many objects, transport, infrastructure, and distance to market are not secondary, but decisive.

TABLE 3.6
Logistics Checklist

Check Point	What to Look At	Warning Sign
Where is the object located relative to processing and market?	Distance to plant, rail, port, customer, trans-shipment point	The project is remote, but this is presented as a secondary detail
Is there access to key infrastructure?	Roads, year-round access, rail, port, inland waterway, power supply	The project cannot work without new infrastructure, but this is understated
What is the transport burden in the economics?	Delivery cost per tonne of raw material or product, share of logistics in netback	Logistics consumes a large part of value, but the model smooths this out
Are there seasonal constraints?	Rain, snow, ice, floods, navigation limits, road condition	The project assumes year-round operations, but logistics is seasonal in reality
Are there multiple transport legs or reloads?	Trucking, rail, port handling, shipping, stockpiling	The chain is long and expensive, but is described too simply
Does the product require special logistics?	Drying, moisture protection, parcel accumulation, blending, special storage	The product requires additional handling, but this is not captured
Are there volume constraints in transport?	Road, rail, station, port, wagon, vessel, or storage capacity	The model assumes a volume that the infrastructure cannot physically support
Is there dependence on a single logistics channel?	One port, one rail station, one route, one contractor	Any disruption in one link puts the entire project at risk
How stable are logistics costs?	Tariffs, fuel, handling, FX risk, freight rates	The project is sensitive to logistics costs, but this is not reflected in the analysis
Are major infrastructure investments required?	Roads, power lines, water supply, terminals, stockpiles	Infrastructure has become a CAPEX project in its own right, but is treated too lightly
Is there a logistics advantage, or a structural disadvantage?	Proximity to market, short haul, water access, integration with existing systems	The object is disadvantaged relative to competitors, but this is not discussed
Is it clear where product ends and commercial delivery begins?	FOB, CIF, EXW Ex Works (EXW), delivery to customer, transfer of risk and price	Price and revenue mix product value with unaccounted logistics burden

3.2.6.1 Practical Conclusion

The key question is whether the project is really a project about the raw material, or in fact a project about distance, infrastructure, and delivery cost. If logistics is clear, accessible, cost-stable, and consistent with project scale, that is a strength. If the transport chain is long, vulnerable, seasonal, or capital-intensive, logistics may be the main project constraint even where geology and processing are acceptable.

3.2.7 Checklist on Revenue and Unit Economics

This checklist is intended to show quickly how the project converts contained value into revenue, and how much value is actually generated per unit of material. This is where it becomes visible whether the project has a real economic basis, or whether it still looks stronger in terms of metal in the ground than in actual money.

TABLE 3.7
Checklist on Revenue and Unit Economics

Check Point	What to Look At	Warning Sign
Is it clear how grade becomes product?	Link between grade, recovery, product yield, and product quality	There is a resource and a grade, but no clear transition to saleable product
Is the payable portion of the valuable component understood?	Payable metal or payable product share, losses, quality constraints	The calculations effectively assume the whole metal or the whole product is paid for
Are real commercial deductions included?	Treatment charges, penalties, discounts, moisture deductions, transport deductions	The model uses price without proper commercial deductions
Is revenue per tonne of feed material understood?	Dollars per tonne of ore, raw material, or mined material, not just total project revenue	The economics exist only in large totals and cannot be read per unit of material
Is it clear what most strongly drives revenue?	Price, recovery, concentrate yield, product quality, impurities	The model does not show which parameter is actually critical
Is contained metal being confused with actual revenue?	Metal in the ground versus metal that is actually paid for	Too much attention is given to in-situ metal, too little to what is paid
Is revenue different across different ore types?	Differences in recovery, payability, product quality, impurities	The model uses one average revenue even though ore types differ fundamentally
Is net revenue per unit of product understood?	Realized price after commercial and transport deductions	The model has a product price, but no clear net revenue
Is there a risk that the product is technically achievable but commercially weak?	Low payability, high penalties, unstable specification	The product can be made, but leaves very little real revenue
Is it clear where the project actually earns its money?	Which stream creates value: quality, product yield, high price, by-products	Revenue exists formally, but the source of the project's economic strength is unclear
Are by-products present, and what is their real role?	By-product metals or products, their contribution to revenue, stability of their market	The project depends too heavily on a by-product for its economics
Does the unit economics look too good?	Comparability with peers, realism of assumptions, revenue structure	The numbers look strong, but only through overly optimistic assumptions

3.2.7.1 Practical Conclusion

The aim is to move away from abstract discussion of grade and large model totals, and towards a harder question: how much revenue does a unit of material actually generate, and from what exactly? If the unit economics are clear, the transition from grade to net revenue is understood, and the key drivers are named honestly, the project can be taken further with more confidence. If the economics depend on heavy averaging, inflated prices, or unclear payability, the revenue side is still weak.

3.2.8 Cost Checklist

This checklist is intended to show quickly where the project is spending money, which cost lines determine its resilience, and whether the economics depend on assumptions that are too optimistic.

3.2.8.1 Practical Conclusion

The key is not just the general cost level, but which items determine the project's fate. If the structure is readable, key cost lines are realistic, and unit costs are clear, the project can be taken further confidently. If costs are too aggregated, key items are understated, or some lines appear to be missing altogether, the cost side may be the main source of weakness.

3.2.9 Sensitivity and Resilience Checklist

This checklist is intended to show quickly how resilient the project is to changes in key assumptions, and whether its economics rest on too narrow a set of favourable conditions.

3.2.9.1 Practical Conclusion

The key question is not merely whether the project is positive in the base case, but how resilient it is to worsening conditions. If it is clear which parameter matters most, where the real limits lie, and whether there is a true buffer, the evaluation becomes meaningful. If the project works only in a "beautiful" model and loses strength at the first movement in assumptions, that is one of the clearest signs of a weak structure.

3.2.10 Checklist on Time and Mining Sequence

This checklist is intended to show quickly how the project distributes value over time, which parts of the deposit support early cash flow, and whether project attractiveness is being created through overly aggressive extraction of the best zones at the front end.

3.2.10.1 Practical Conclusion

The key question is not only how much value the project contains in total, but when that value appears. If the early years are strong without artificially "eating" the best zones, and the later part of the project remains economically meaningful, the time

TABLE 3.8
Cost Checklist

Check Point	What to Look At	Warning Sign
Is the main source of cost understood?	Mining, stripping, processing, power, water, reagents, transport, tailings, infrastructure	Costs are presented in aggregate, but the main driver is unclear
Is the cost structure visible by major stage?	Mining, processing, logistics, overheads separately	The economics are condensed into one averaged number without structure
Which costs are imposed by the nature of the object?	Depth, strip ratio, rock hardness, water conditions, remoteness, ore properties	The project presents costs as "normal" even though the object is clearly difficult
Which costs depend on the chosen scheme?	Mining method, grind size, process complexity, selectivity, logistics	A weak scheme is disguised as unavoidable geological reality
Are any cost lines suspiciously low?	Power, fuel, reagents, transport, maintenance, water, tailings	One or more key items is clearly understated relative to peers
Are costs of meeting product quality included?	Blending, sorting, drying, quality control, upgrading	The model has a saleable product, but the cost of getting there is missing
Are infrastructure and support costs included?	Roads, power, water, workshops, storage	The production route exists, but the support system appears to be "free"
Are logistics costs understated?	Transport, trans-shipment, storage, moisture loss, parcel accumulation	Logistics is discussed separately, but barely felt in the cost structure
Is there sensitivity to the main cost items?	Fuel, power, reagents, spares, contractors, tariffs	The project works only under "ideal" cost conditions
Are unit costs per tonne of material and product clear?	Dollars per tonne of ore, raw material, concentrate, or metal	There is a total cost figure, but unit economics cannot be read
Is processing cost understated?	Grinding, flotation, beneficiation, hydrometallurgy, complex treatment	A technically complex object is shown as cheap to process
Are there cost lines that often appear later?	Sustaining capital, maintenance, equipment replacement, environmental compliance, tailings, dewatering	The early model looks "clean," but too many costs are left outside it

profile looks healthy. If the project is attractive mainly in the beginning while the later profile is much weaker, that should be treated as a risk, not a neutral feature of the model.

3.2.11 Project Maturity Checklist

This checklist is intended to show quickly what stage the project is really at, which key questions are already confirmed, and which still remain assumptions. Its purpose is not to assign a formal stage label, but to avoid confusing project maturity with the volume of material prepared.

TABLE 3.9
Sensitivity and Resilience Checklist

Check Point	What to Look At	Warning Sign
Is it clear which parameter is most sensitive?	Price, recovery, product yield, raw material quality, transport, power, strip ratio, throughput	The project contains many numbers, but it is unclear what actually determines resilience
What happens if price falls?	Changes in revenue, margin, cut-off grade, economic shell	Even a modest price decline quickly destroys the economics
What happens if costs rise?	Fuel, power, reagents, transport, maintenance, contractors	The project works only at an unrealistically "clean" cost level
What happens if recovery or product yield falls?	Effect on revenue, product quality, margin	The economics weaken sharply even under moderate deterioration in processing performance
What happens if feed quality worsens?	Impurities, ore-type variability, ash, moisture, lower grade	The model uses averaged material, but does not show what happens with weaker feed
Is there a resilience buffer?	How far the project is from losing profitability	The project is positive only within a very narrow range of assumptions
Is project behaviour understood under bad, not only base-case, conditions?	Price decline, cost inflation, delayed start-up, weaker logistics	Only the "base case" is shown, without serious testing of weak scenarios
Do parameters move one at a time or together?	Price and logistics, quality and recovery, throughput and cost	Sensitivity is shown artificially one variable at a time, even though risks come in combinations
Is there one "break" factor?	The parameter whose movement causes the economics to fail first	The main limiting factor is not identified or is hidden in the overall model
What happens to early cash flow if conditions deteriorate?	Early years, capital payback, start-up resilience	The project is especially vulnerable early on, but this is not highlighted
Is the project resilient through time?	Not just aggregate economics, but timing of results by year	The project looks strong in total, but weak in time profile
Are the project's true limits understood?	Under which assumptions it remains strong, marginal, or weak	The model says "the project is good," but does not show the boundaries of that conclusion

3.2.11.1 Practical Conclusion

The key question is not whether the project is "liked," but whether it is mature enough for the next decision. If it is clear what has been confirmed, what remains open, and which questions must be closed before the next step, the project is moving sensibly. If maturity is overstated and assumptions are presented as confirmed, that is one of the most reliable early signs of future problems.

3.2.12 Project Financeability Checklist

This checklist is intended to show quickly whether the project can not only look technically and economically convincing, but also be seen as a structure fit for a

TABLE 3.10
Time & Mining Sequence Checklist

Check Point	What to Look At	Warning Sign
Is it clear which zones support the first years?	Ore quality, strip ratio, accessibility, processing simplicity, start-up logistics	The early years look strong, but it is unclear why
Is the project overly dependent on bringing the best zones forward?	High grades, easy ore, best process zones early in the schedule	Early attractiveness is achieved by accelerated consumption of the best material
What happens to material quality later?	Lower grades, higher impurities, more difficult ore types, worse processing	Strong early indicators conceal a weak later-stage project
How do strip ratio or mining complexity change over time?	Depth increase, geometry deterioration, longer hauls, worsening mining conditions	Early years are easy, later years much harder, but this is weakly reflected
Is the project's strength an illusion created by the first years?	Early cash flow, early margin, early revenue	The model looks convincing mainly because the start has been artificially strengthened
Does the later part of the project retain value?	Material quality, costs, revenue, economic role of late years	A long mine life adds little real value, but is presented as an advantage
Is the link between sequence and resilience clear?	Which zones provide a buffer, which make the project vulnerable	There is a mine schedule, but its economic significance is not explained
Is later-stage economics being weakened for the sake of a strong start?	Pulling best zones early, weakening the later profile	Early years are improved, but the later project is materially weakened
Is there a strong early project core?	Areas that genuinely support capital recovery and start-up resilience	All the good material lies too far out, and the start is weak
Is the project critically dependent on success in the first years?	Capital payback, debt service, resilience to weak markets	Any early deterioration sharply weakens the entire project
Is the link between mining schedule and feed quality understood?	Feed consistency, blending, transitions between ore types	Sequence is shown technically, but not tied to processing and product
Is the production schedule more than a formality?	Whether it has economic logic, not just extraction order	The schedule exists, but it is unclear why this sequence was chosen

serious capital discussion. Its purpose is not to "approve" financing in advance, but to test whether the project is genuinely ready for the expectations of an investor, lender, strategic partner, or product buyer.

3.2.12.1 Practical Conclusion

The key question is whether the project is merely an interesting object, or whether it can already be understood as a structure legible to capital. If the source of cash

TABLE 3.11
Project Maturity Checklist

Check Point	What to Look At	Warning Sign
Is it clear what stage the project is really at?	Possibility, concept, preliminary evaluation, more mature development, investment readiness	The project is presented as more mature than its factual basis allows
What has already been confirmed, and what is still assumed?	Geology, processing, product, market, logistics, costs, schedule	Assumptions are presented as proven facts
Have the key geological uncertainties for this stage been closed out?	Data density, model, homogeneity, distribution of ore types	Geology is still immature, but strong conclusions are already being drawn
Is the process route confirmed to an adequate level?	Testwork representativeness, feed variability, stability of results	The technology is described confidently, but the test base is weak
Is the product and market understood at a level sufficient for the stage?	Specification, buyers, pricing, commercial constraints	The commercial side remains too abstract
Are costs and capital estimates worked up realistically?	Cost structure, depth of analysis, major remaining uncertainties	The model already contains "precise" numbers at a very early stage
Is it clear which key risks are still open?	Geology, metallurgy, logistics, environment, infrastructure, market	The project contains a lot of optimism, but too little honest discussion of open risks
Does the level of work support the decisions being made on its basis?	Whether the depth of analysis matches the weight of the decision	A very serious next step is being justified on a weak base
Is concept being confused with readiness for execution?	Difference between "one can imagine it" and "one can build it"	The project looks good as an idea, but is presented as nearly execution-ready
Is there one critical element that is still unconfirmed?	Water, power, tailings, recovery, market, infrastructure, permits	One supporting element remains weak, but this is being hidden
Is it clear what must be proven before the next stage?	Specific tasks for the next step, not just general wishes	The project is "moving forward," but without a clear idea of what still needs to be proved
Has the project remained too long with the same uncertainty unresolved?	Repeating weak points from stage to stage	The same open questions persist for years while the project continues advancing

flow is clear, key assumptions are confirmed, the project stage matches the type of capital being sought, and the project can withstand hard questions, that indicates a high degree of readiness.

If, by contrast, the project rests on optimism, responds poorly to doubt, and does not understand who should enter it or why, then its financeability remains weak for now.

TABLE 3.12
Project Financeability Checklist

Check Point	What to Look At	Warning Sign
Is it clear how the project returns capital?	Source of cash flow, early years, margin resilience, payback period	The project looks interesting, but it is unclear how it actually returns money
Does the project have a sufficient resilience buffer?	Behaviour under lower prices, higher costs, weaker recovery, delayed start-up	The project works only under overly favourable assumptions
Are the key assumptions confirmed?	Geology, processing, product, market, logistics, costs, schedule	Too much of the project still rests on assumptions
Is it clear who could finance the project at this stage?	Early-stage capital, strategic investor, product buyer, debt, mixed structure	The project is seeking "any money," without understanding what kind of capital fits
Does the project stage match the type of capital being targeted?	Whether maturity matches investor or lender expectations	A very immature project is trying to talk to the most demanding capital
Is the main risk clear from the capital point of view?	Geology, technology, product, market, country, logistics, schedule, CAPEX	The team speaks about upside, but cannot clearly name the main risk
Is there a clear logic of risk and reward?	Relationship between potential return and uncertainty	The project is risky, but offers too little reason to accept that risk
Is it clear why the project should matter to an external party?	Return of capital, access to raw material, control over supply, strategic importance, value growth	The project seems attractive only to its own sponsors
Can the project withstand hard questions?	What happens if price falls, costs rise, logistics weaken, quality deteriorates	The project logic starts to fall apart after the first serious question
Is the project internally coherent?	Link between geology, technology, product, market, costs, and cash flow	The elements look individually acceptable, but do not form one system
Is it clear what the money is needed for now?	Drilling, testwork, engineering, infrastructure, construction, start-up, working capital	Capital is being requested, but it is unclear which stage it is supposed to fund
Is the project trying to look more financeable than it really is?	Overstated maturity, excessive optimism, understated risk, "beautiful packaging"	The external presentation is stronger than the project's real readiness

3.3 KEY DUE DILIGENCE QUESTIONS

In mining, due diligence should never be reduced either to a formal document check or to the mechanical collection of large volumes of material. Its practical purpose is different: to test how reliable the assumptions are on which the value of the project or asset is built.

In other words, the purpose of due diligence is not to gather the maximum amount of information, but to understand which factors actually determine the future viability of the object and how well those factors are truly supported.

That is why due diligence always depends on the purpose of the review. For a buyer, one of the main questions will be the real value of the asset, its hidden risks, and its ability to generate cash flow. For a lender, what matters more is project resilience, the reliability of the assumptions, and the ability to service obligations. For a strategic partner, the priority may shift towards access to raw material, product quality, the asset's place in the supply chain, and the possibility of integration. For a seller, the critical question is whether the project can withstand external professional scepticism and not fall apart under the first serious test.

That is why the first due diligence question should not be: "What else can be checked?" It should be: "Which assumptions need to be confirmed in order to make the decision without self-deception?"

3.3.1 What Decision is the Due Diligence Actually for?

Before any review begins, it is necessary to define clearly what decision is supposed to be made on the basis of it. Without that, due diligence quickly turns into an endless list of questions, many of which are either secondary or irrelevant to the outcome.

If the review concerns the acquisition of an early-stage project, the priority will shift towards the geological foundation, data quality, technological achievability of the product, the logic of future work, and the realism of the assumptions on which value rests. If the review concerns a more mature project being prepared for financing or sale, the focus shifts towards the resilience of cash flow, sensitivity to market change, confirmation of the process route, realism of capital and operating costs, and the project's ability to withstand adverse scenarios. If an operating asset is being reviewed, the critical issues become actual operating performance, product quality, equipment condition, hidden liabilities, future mining complications, contracts, and the real behaviour of costs.

That is why, before the work begins, several questions must be answered:

What decision is to be taken on the basis of the review?
Which assumptions are decisive for that decision?
Which of them are already sufficiently confirmed, and which still exist mainly at the level of model, expectation, or presentation?
Which type of risk matters most here: geological, technological, market, logistics, financial, legal, regulatory, or stage-of-maturity risk?

Until those questions are asked, the review almost inevitably begins to sprawl and lose precision.

3.3.2 What Exactly is Being Reviewed: a Licence, a Project, or an Operating Asset?

One of the most common mistakes is to evaluate different types of objects in the same way. In reality, due diligence for an early-stage licence, for a development-stage project, and for an operating asset should be built differently.

If the object is mainly a licence and a mineral base, then the review should focus on subsurface rights, access to the ground, the geological base, data quality, technological feasibility, the market for the future product, logistics, and the project's general ability to move to the next stage. In that case, due diligence is testing whether the licence can actually be turned into a real project.

If, however, the object is an operating asset, another layer of review must be added. In that case, it is necessary to evaluate not only the mineral base and the project logic, but the whole production system. Questions arise about the actual condition of the pit or mine, the plant, the tailings system, stockyards, workshops, power supply, water supply, roads, rail infrastructure, port capacity, and auxiliary facilities. Rolling stock, mining equipment, process equipment, physical wear, remaining service life, repair history, replacement needs, and whether actual performance matches stated capacity must all be checked separately.

In other words, if the object already exists as an enterprise, due diligence must answer not only the question of how much ore or product it contains, but also the question of what condition the production machine itself is in, the machine that must mine the ore, process it, and turn it into money.

This is a fundamental point. The same geologically strong asset may have very different value depending on whether it is simply a licence with future potential, or an operating business with ready but worn and capital-hungry infrastructure.

That is why, at the outset of due diligence, it must be established clearly what exactly is being evaluated:

only the mineral right,
a project under development,
or an operating asset with physical infrastructure, equipment, personnel, and operating history.

Without that, it is impossible to set the right priorities for the review.

3.3.3 Is There Real Control Over the Asset and Access to It?

Once the type of object is defined, the next task is to verify whether it is actually controlled under conditions that allow it to be developed.

In practice, it is necessary to verify not only the licence in a formal sense, but the whole system of control around the asset.

Who exactly holds the subsurface right?
On what conditions is that right held?
What is its duration?
Are there obligations regarding minimum work, reporting, investment, or development timing?
Is separate approval required in the event of a change of ownership?
Are there encumbrances, royalties, rights of third parties, obligations to previous owners, supply commitments, land restrictions, or other factors that could weaken control over the asset?

No less important is the issue of access to surface rights and infrastructure. Even if the mineral right exists, that does not mean the project can freely build roads, power lines, tailings facilities, water intakes, plant sites, and other structures without which mining and processing are impossible.

A good due diligence review in this area does not ask the formal question "is there a licence?" It asks the practical question: is the asset really controlled in the legal and spatial form required to turn it into a working project, or to maintain it as a working asset?

3.3.4 How Reliable is the Geological and Resource Foundation?

The next level of review concerns the reliability of the geological base on which the value of the asset is built.

If the asset is an early-stage project, the focus should be on data quality, drill density and distribution, sample representativeness, analytical QA/QC, geological model logic, reasonableness of interpolation, deposit type, mineralization continuity, and the distribution of ore types.

It is necessary to review not only the final resource or reserve figure, but the path by which it was produced.

What is the data density?
How representative is the sampling?
Is there a coherent geological model?
Does the level of interpolation match the actual level of knowledge?
Does the model create a false sense of precision?
Is the project's attractiveness resting on zones that are less well supported than the external presentation suggests?

If the object is operating, another question is added: to what extent does the actual behaviour of the deposit confirm or contradict the model? Do real losses, dilution, feed quality, orebody geometry, and the distribution of ore types match what was built into the project basis? If not, it is necessary to understand whether this is a local operating issue or a symptom of a deeper geological underestimation of the object's complexity.

What matters here is the difference between neat presentation and real confirmation. A strong 3D model, an attractive block model, and precise numbers may all look convincing. But due diligence must test not visual persuasiveness, but the strength of the foundation.

3.3.5 Does Raw Material Quality and Product Quality Support the Stated Economics?

One of the most common weaknesses in poor due diligence is that volume is assessed before quality, and grade before saleable outcome.

In reality, it is often raw material and product quality that determine whether value survives in the project or disappears on the way to market.

Due diligence in this area must test not only average values, but also their variability, their distribution through the mining sequence, their dependence on ore type, and their effect on both technology and marketability.

For iron ore projects, this means that Fe grade alone is not enough. Phosphorus, sulphur, silica, alumina, ore type, moisture, size distribution, and the achievable final product specification must all be reviewed.

For copper projects, it is important to understand not only copper grade, but also the character of the mineralization. How homogeneous is the sulphide component? Are there oxide and mixed ores? How do they affect recovery, flotation, concentrate quality, and the need for separate process streams?

For gold projects, due diligence must almost always ask about refractoriness. Are there carbonaceous components? Is there preg-robbing? What is the role of arsenopyrite, pyrite, and other minerals in which the gold is finely locked? Is pre-treatment required through roasting, pressure oxidation, bio-oxidation, or some other capital-intensive step?

For coal and industrial minerals, the focus must be not only on general quality parameters, but on their practical commercial meaning. What are the ash, moisture, sulphur, saleable yield, quality consistency, chemical purity, whiteness, reactivity, suitability for use, and price sensitivity to deviations in these properties?

A strong due diligence review in this block always tests the same thing: whether grade is being confused with value.

3.3.6 Is the Process Route Confirmed, and How Robust Is It?

Even with a strong geological base and good raw material quality, a project can fail at the processing level. The next due diligence block therefore concerns processing and technological feasibility.

The key question here is not whether a technology exists "in principle," but whether there is a realistic, confirmed, and economically viable process route for this specific material.

The representativeness of the testwork must be checked. On what material was it performed? Does it cover all key ore types? Does it capture quality variability? Were good results obtained only on the easiest samples?

The next question is how sensitive the technology is to feed quality. Many projects look good on averaged material and weaken sharply when mineralogy, impurities, particle size, moisture, or the share of transitional ore changes.

It is also necessary to test whether acceptable recovery depends on a route that is excessively expensive, energy-intensive, or operationally complex. Sometimes the problem is not that metallurgy is impossible, but that it is too heavy in CAPEX, operating cost, reagents, power, or scale requirements.

For an operating asset, this area requires an additional level of review. One must examine not only the design flowsheet, but the actual behaviour of the plant or process unit.

Does actual recovery match the stated recovery?
How stable is performance across shifts and time periods?

Are there throughput bottlenecks?
What is the actual condition of the equipment?
Where are the persistent bottlenecks?
Is there accumulated technical wear that is not yet reflected in value?

If due diligence does not test how the technology performs on real feed rather than idealized feed, it is not reviewing the project itself. It is reviewing only its best version.

3.3.7 What Does Actual Performance Show, if the Asset is Already Operating?

For an operating asset, due diligence has a special role. In this case, it becomes possible to test not only the model, but the actual behaviour of the whole system.

The critical items here are:

- mining history,
- actual feed quality,
- recovery,
- saleable product yield,
- plant throughput,
- real costs,
- equipment reliability,
- infrastructure condition,
- incident history,
- repair history,
- dependence on particular contractors or key staff,
- and the quality of operational management.

It is very important to compare what the asset promises with what it actually does.

Do actual mining volumes match stated levels?
How do losses and dilution behave?
How stable is product quality?
What are the real mining, processing, and logistics costs?
How reliable are the key machines and units?
What is the actual remaining value of the equipment, and how much capital will be required to keep it operational?
Are there large future costs hidden behind a seemingly functioning operation, replacement of equipment, major repairs, plant rebuilds, tailings strengthening, or infrastructure renewal?

This is where due diligence must recognize that an operating asset consists not only of the licence and the mineral base, but also of buildings, structures, mobile equipment, transport, process equipment, power systems, workshops, warehouses, administration, and the whole support system required for production. All of this must be treated not merely as "existing infrastructure," but as assets with a specific condition, wear profile, remaining life, and future capital requirement.

In some cases, an operation looks stronger than it really is simply because current production is still running on old equipment and infrastructure that are already approaching the point of major reinvestment. This is one of the classic risks of superficial due diligence.

3.3.8 Which Future Risks Must Not Be Missed?

Asset value review cannot be limited to current condition. It must also test future factors capable of changing operating conditions, costs, permitting status, environmental obligations, logistics resilience, and market position.

These include:

- environmental obligations,
- future rehabilitation and closure costs,
- tailings management,
- water balance,
- dam stability,
- the need for infrastructure expansion,
- dependence on external power or water,
- increasing mining depth and complexity,
- rising strip ratio,
- declining material quality over time,
- labour and social risks,
- possible changes in regulation,
- permitting constraints,
- and potential points of conflict with local communities or land users.

What is especially important is that many of these factors are not current problems. Weak due diligence therefore misses them easily. Strong value review, however, must look forward. The question is not only what exists today, but what will first begin to weaken the project as it develops.

3.3.9 Do Market, Contracts, and Logistics Actually Support the Project?

As in the earlier sections, due diligence must test not merely whether a product exists, but whether there is a path from product to money.

In practice, this means checking several levels.

How exactly is the product sold?

Is this an exchange, index, contract, regional, or niche market?

Is there a real buyer or buyer group for the product quality the project can actually deliver, not for an idealized product?

How is price formed?

What are the commercial deductions, penalties, discounts, requirements for volume, quality, moisture, size distribution, and consistency of supply?
What is the role of logistics in net revenue?
Is there access to rail, port, roads, all-season routes, water, and power?
Has transport itself become a project inside the project?

For an operating asset, due diligence should also review the actual contracts in place, the quality of their terms, dependence on particular buyers, duration of relationships, pricing mechanisms, and the real risks around renewal or replacement of those arrangements.

3.3.10 Is Value Being Overstated through Weak Assumptions?

In the end, due diligence almost always comes back to one main question: is the stated value the result of overly optimistic assumptions, a weak model structure, or a poor understanding of risk?

This is where the most uncomfortable questions must be asked.

Is the assumed price too high?
Are costs understated?
Are ore types, recovery, and product quality over-averaged?
Does logistics look simpler than it really is?
Has early economics been improved through an overly aggressive mining sequence?
Is gross product price being substituted for real payable revenue?
Are closure, environmental, sustaining capital, repair, and infrastructure obligations understated?
Is the condition of existing buildings, equipment, and production capacity being viewed too optimistically?
Is there a hidden need for major future capital that is not yet reflected in the model?

It is very important to understand that due diligence tests not only individual numbers, but their internal coherence. A strong model is not simply one in which the figures are neat. It is one in which geology, product, market, cost, time, infrastructure, and capital are aligned and do not contradict one another.

If a project shows attractive value only under a combination of several favourable assumptions at once, that is already a due diligence finding.

3.3.11 Where Does Due Diligence Most Often Find Weakness?

Practice shows that the main weaknesses found in due diligence usually sit not in one isolated place, but in the transitions between disciplines.

Between legal title and the practical ability to develop the asset.
Between geology and resource estimation.

Between the resource and raw material quality.
Between raw material and processing.
Between processing and saleable product.
Between product and market.
Between revenue and net cash result.
Between the model and the actual behaviour of an operating asset.
Between technical attractiveness and financeability.

That is why strong due diligence must not be merely a list of sections. It must be a test of project coherence as a system.

Very often, the value of an asset weakens not because the geology is poor and not because the product is impossible, but because somewhere between those blocks there is a disconnect that the model hides and reality later exposes.

3.3.11.1 Practical Conclusion

The purpose of due diligence in mining is not to confirm as many strengths of the asset as possible. Its task is to identify the assumptions on which the project's future truly depends, and to test how robust those assumptions are under an external, professionally skeptical view.

Good due diligence does not aim to collect the maximum amount of material. It aims to understand where the logic of the project may break.

Is the asset truly controlled?
Is the geological base reliable?
Is the product real?
Is the process route confirmed?
If the asset is already operating, what is the condition of its infrastructure, buildings, mining fleet, and process equipment?
Do market and logistics actually support the project?
Are future risks understood?
And is value being overstated through assumptions that cannot withstand serious review?

If there are clear and supported answers to those questions, due diligence becomes a tool for sound decision-making. If the answers are vague and the project rests on an attractive but fragile structure, then due diligence should reveal that before an expensive and hard-to-reverse decision is made.

3.4 PRACTICAL CALCULATIONS AND APPLIED CASES

In the previous sections, the general logic of evaluation was laid out step by step: how a geological object gradually becomes a product, revenue, cash flow, and, ultimately, a project that may or may not withstand economic and investment scrutiny. In real work, however, general logic alone is not enough. At some point, one still has to open a calculator, Excel, or Python and look at what actually stands behind the number in the table.

This section brings together simple practical calculations and applied cases that help test the order of magnitude quickly, compare alternatives, and understand how grade, recovery, price, cost, product quality, logistics, and time are converted into economic result. The purpose is not to replace a full model with a few formulas. The purpose is more practical than that: to provide working tools for a rapid test of project logic.

Such calculations are especially useful at first contact with an object, during internal discussion, when comparing projects, in due diligence, and whenever it is necessary to understand quickly how robust a particular assumption really is. That is why this section is often especially close to those who like to run the numbers by hand and are not willing to accept any figure on trust simply because it already appears in a model, spreadsheet, or presentation.

At the same time, a few simple points should be kept in mind.

First, neither metal equivalent, nor cut-off grade, economic threshold, or revenue per tonne of material is a permanent characteristic of a deposit. All of these values depend on the timing of the calculation and on the assumptions adopted at that point: cut-off parameters, prices, costs, recovery, payable terms, product quality, logistics, and process route.

Second, in calculations the problem often arises not because the formula itself is wrong, but because the value inserted into it does not actually belong to the conditions being analyzed. The formula may be entirely correct, but the price may be taken for the wrong product, the recovery for the wrong ore type, the cost for the wrong project stage, or the averaged value for a reality different from the one the model is trying to describe.

In other words, it is not necessarily the mathematics that fails. More often, it is the person who was too confident that a variable that looked appropriate was also appropriate in meaning.

3.4.1 How to Calculate Metal Equivalent

Metal equivalent is used to convert the combined value of several components into one notional indicator. This is convenient for rapid evaluation and comparison, but one point must always be kept in mind: metal equivalent always depends on the timing of the calculation, prices, recoveries, and the other assumptions adopted. It is not a permanent characteristic of the ore, but a calculated value.

As an example, take a polymetallic ore with the following grades and recoveries:

oxidized copper 0.45%,
sulphide copper 0.80%,
gold 0.65 g/t,
silver 11 g/t;
and recoveries of 66%, 83%, 75%, and 67%, respectively.

For simplicity, let us use rounded prices:

copper USD 13,000/t,
gold USD 4,800/oz,
silver USD 84/oz.

3.4.2 General Logic of the Calculation

The first step is to calculate the value contribution of each component in 1 tonne of ore. After that, the total value of all components can be converted either into a gold equivalent or into a copper equivalent.

3.4.3 Value of a Component in 1 Tonne of Ore

For a metal whose grade is expressed in percent, the formula is:

Value of component in 1 tonne of ore = grade, fraction × recovery, fraction × metal price, USD/t

For a metal whose grade is expressed in g/t, the formula is:

Value of component in 1 tonne of ore = grade, g/t × recovery, fraction × metal price, USD/oz / 31.1035 g/oz

where 31.1035 g/oz is the number of grams in one troy ounce.

3.4.4 Calculating the Contribution of Each Component

3.4.4.1 Oxidized Copper

Formula:

Value of oxidized copper = Cu(ox), fraction × recovery, fraction × copper price, USD/t

Substitution:

0.45% × 66% × USD 13,000/t = 0.0045 t/t × 0.66 × USD 13,000/t Cu = USD 38.6/t ore

Rounded:

USD 39/t ore

3.4.4.2 Sulphide Copper

Formula:

Value of sulphide copper = Cu(sulph), fraction × recovery, fraction × copper price, USD/t

Substitution:

0.80% × 83% × USD 13,000/t = 0.0080 t/t × 0.83 × USD 13,000/t Cu = USD 86.3/t ore

Rounded:

USD 86/t ore

3.4.4.3 Gold

Formula:

Value of gold = Au, g/t × recovery, fraction × gold price, USD/oz / 31.1035 g/oz

Substitution:

0.65 g/t × 75% × USD 4,800/oz / 31.1035 g/oz = 0.65 × 0.75 × USD 4,800 / 31.1035 = USD 75.2/t ore

Rounded:

USD 75/t ore

3.4.4.4 Silver

Formula:

Value of silver = Ag, g/t × recovery, fraction × silver price, USD/oz / 31.1035 g/oz

Substitution:

11 g/t × 67% × USD 84/oz / 31.1035 g/oz = 11 × 0.67 × USD 84 / 31.1035 = USD 19.9/t ore

Rounded:

USD 20/t ore

3.4.5 Total Value of All Components

Formula:

Total value = value of Cu(ox) + value of Cu(sulph) + value of Au + value of Ag

Substitution:

USD 39/t ore + USD 86/t ore + USD 75/t ore + USD 20/t ore = **USD 220/t ore**

3.4.6 Calculating Gold Equivalent

First, determine the value of 1 g Au in ore at 75% recovery.

Formula:

Value of 1 g Au in ore = gold price, USD/oz / 31.1035 g/oz × Au recovery, fraction

Substitution:

USD 4,800/oz / 31.1035 g/oz × 0.75 = USD 115.7/g Au in ore

Rounded:

USD 116/g Au in ore

Gold equivalent can now be calculated.
Formula:

AuEq = total ore value, USD/t / value of 1 g Au in ore, USD/g

Substitution:

USD 220/t ore / USD 116/g Au = **1.9 g/t AuEq**

Result:

AuEq ≈ 1.9 g/t

3.4.7 Calculating Copper Equivalent

This is where care is needed. Since the ore contains oxidized copper and sulphide copper with different recoveries, the first step is to calculate a weighted average copper recovery by metal share.

Assume:

35% oxidized copper, that is 0.35,
65% sulphide copper, that is 0.65.

3.4.7.1 Weighted Average Copper Recovery

Formula:

Recovery Cu(avg) = share of oxidized Cu × recovery of oxidized Cu + share of sulphide Cu × recovery of sulphide Cu

Substitution:

$0.35 \times 0.66 + 0.65 \times 0.83 = 0.231 + 0.540 = \mathbf{0.771}$

Or:

77.1%

Rounded:

77%

3.4.7.2 Value of 1% Cu in Ore

Formula:

Value of 1% Cu in ore = 0.01 t/t × copper price, USD/t × average Cu recovery

Substitution:

0.01 × USD 13,000/t Cu × 0.771 = **USD 100.2/t ore for each 1% Cu**

Rounded:

USD 100/t ore for each 1% Cu

3.4.7.3 Copper Equivalent

Formula:

CuEq = total ore value, USD/t / value of 1% Cu in ore, USD/t per 1% Cu

Substitution:

USD 220/t ore / USD 100/t ore per 1% Cu = **2.2% CuEq**

3.4.8 Result

CuEq ≈ 2.2%

3.4.9 Summary

For this example, at the assumed prices, the results are:

AuEq ≈ 1.9 g/t
CuEq ≈ 2.2%

3.4.10 What Matters Here

The formula may be correct, and the result may still be weak in meaning if the wrong variables have been inserted into it.

The price may refer to the wrong product.
The recovery may have been taken for the wrong ore type.
Averaging may have been done without regard to the real structure of the material.
And the prices themselves may belong to a different market situation.

That is why, when calculating an equivalent, one must watch not only the formula, but also the applicability of every input variable. The error usually begins not in the mathematics, but slightly earlier.

Metal equivalent is useful as a rapid working indicator, but it should be used with care. It depends not only on grades, but also on recovery, and therefore is already tied to a specific process route. It also depends on prices, and therefore on the timing of the calculation and the market environment. Metal equivalent should never be treated as a permanent characteristic of the ore. It is not a geological constant, but a calculated quantity sensitive to both technology and time.

3.4.11 How to Calculate Cut-Off Grade

Cut-off grade shows the minimum grade of the valuable component, or the minimum saleable product yield, at which 1 tonne of ore still covers its costs.

For single-component ore, the calculation is usually more straightforward. For polymetallic ore, or for iron ore where the concentrate yield is decisive, the calculation is less direct. But the basic logic is always the same: the value of a unit of material must be compared with the costs carried by that unit.

Below are three simple examples: copper, gold, and magnetite iron ore.

3.4.11.1 Example for Copper Ore

Assume:

copper price = **USD 13,000/t Cu**
overall recovery = **0.77 t Cu/t Cu in ore**, or **77%**
total cost = **USD 30/t ore**
First, determine the value of 1% Cu in 1 tonne of ore.

Formula:

Value of 1% Cu in 1 tonne of ore = 0.01 t Cu/t ore × Cu price, USD/t Cu × recovery

Substitution:

0.01 × USD 13,000 × 0.77 = **USD 100.1/t ore**

Rounded:

USD 100/t ore for each 1% Cu

Now calculate cut-off grade.
Formula:

Cut-off grade, % Cu = cost, USD/t ore / value of 1% Cu in 1 tonne of ore, USD/t ore

Substitution:

USD 30/t ore / USD 100/t ore per 1% Cu = **0.30% Cu**

3.4.11.2 Result

At these assumptions, the cut-off grade is:

0.30% Cu

This means that at a copper price of USD 13,000/t Cu, a recovery of 0.77, and a cost of USD 30/t ore, material grading about 0.30% Cu lies at the economic margin.

3.4.11.3 Example for Gold-Bearing Ore

Now consider a single-component gold ore.

Assume:

gold price = **USD 4,800/oz Au**
overall recovery = **0.75 t Au/t Au in ore**, or **75%**
total cost = **USD 50/t ore**

First, determine the value of 1 g Au in 1 tonne of ore.
Formula:

Value of 1 g Au in 1 tonne of ore = gold price, USD/oz / 31.1035 g/oz × recovery

Substitution:

USD 4,800/oz / 31.1035 × 0.75 = **USD 115.7/g Au in 1 tonne of ore**

Rounded:

USD 116/g Au in 1 tonne of ore

Now calculate cut-off grade.
Formula:

Cut-off grade, g Au/t ore = cost, USD/t ore / value of 1 g Au in 1 tonne of ore, USD/g Au

Substitution:

USD 50/t ore / USD 116/g Au = **0.43 g Au/t ore**

3.4.11.4 Result

At these assumptions, the cut-off grade is:

0.43 g Au/t ore

This means that at a gold price of USD 4,800/oz Au, a recovery of 0.75, and a cost of USD 50/t ore, material grading about 0.43 g Au/t ore lies at the economic margin.

3.4.11.5 Example for Magnetite Iron Ore

For iron ore projects, cut-off grade is often more usefully considered not directly through Fe grade, but through the ore's ability to produce a saleable concentrate.

Assume the following inputs:

62% Fe concentrate price FOB = **USD 91/t concentrate**
mining cost = **USD 3/t ore**
beneficiation cost = **USD 15/t ore**
concentrate transport to port = **USD 10/t concentrate**
concentrate yield = **0.35 t concentrate/t ore** at **35%** **Fe** feed
concentrate yield = **0.10 t concentrate/t ore** at **15%** **Fe** feed

Step 1. Determine the net value of 1 tonne of concentrate
Formula:

Net value of 1 tonne of concentrate = concentrate price, USD/t concentrate − transport, USD/t concentrate

Substitution:

USD 91/t concentrate − USD 10/t concentrate = **USD 81/t concentrate**

Step 2. Determine the cost per tonne of ore
Formula:

Cost per tonne of ore = mining, USD/t ore + beneficiation, USD/t ore

Substitution:

USD 3/t ore + USD 15/t ore = **USD 18/t ore**

Step 3. Find the critical concentrate yield
At the economic margin, concentrate revenue must exactly cover ore cost.
Formula:

Critical concentrate yield, t concentrate/t ore = cost per tonne of ore, USD/t ore / net value of 1 tonne of concentrate, USD/t concentrate

Substitution:

USD 18/t ore / USD 81/t concentrate = **0.222 t concentrate/t ore**

Or:

22.2% concentrate by mass relative to ore mass

This means that the cut-off boundary lies where 1 tonne of ore yields about 0.222 tonnes of concentrate.

Step 4. Link concentrate yield to Fe grade
Now use the stated relationship between Fe grade in ore and concentrate yield. We have two points:

at 35% Fe, yield = 0.35 t concentrate/t ore
at 15% Fe, yield = 0.10 t concentrate/t ore

For simplicity, assume a linear relationship between these points.
Formula of the straight line:

Concentrate yield, t concentrate/t ore = a × Fe grade, % + b

First, calculate the slope:

$$a = (0.35 - 0.10) / (35 - 15)$$

Substitution:

a = 0.25 / 20 = **0.0125 t concentrate/t ore for each 1% Fe**

Now calculate the intercept b.
Formula:

$$0.35 = 0.0125 \times 35 + b$$

Substitution:

$$0.35 = 0.4375 + b$$

Therefore:

b = **−0.0875**

So the relationship becomes:

Concentrate yield, t concentrate/t ore = 0.0125 × Fe grade, % − 0.0875

Step 5. Insert the critical yield
Now insert the previously calculated cut-off yield of 0.222 t concentrate/t ore.
Formula:

0.222 = 0.0125 × Fe grade, % − 0.0875

Substitution:

$$0.222 + 0.0875 = 0.0125 \times \text{Fe grade}$$
$$0.3095 = 0.0125 \times \text{Fe grade}$$

Now calculate Fe grade:

Fe grade = 0.3095 / 0.0125 = **24.76% Fe**

Rounded:

25% Fe in ore

3.4.11.6 Result

At the assumed inputs, the cut-off grade for this magnetite iron ore is approximately:

25% Fe in ore

That is, at an in-situ grade of about 25% Fe, the material lies at the economic margin if:

- The concentrate sells at USD 91/t FOB.
- Transport to port is USD 10/t concentrate.
- Mining and beneficiation together cost USD 18/t ore.
- And concentrate yield varies from 0.35 t/t at 35% Fe to 0.10 t/t at 15% Fe.

3.4.11.7 What Matters Here?

For copper and gold, cut-off grade was calculated directly because in both cases a single-component ore and constant recovery were assumed.

For iron ore, however, the calculation goes through concentrate yield, t concentrate/t ore, rather than metal grade alone. That is why in iron ore projects, the cut-off grade often has to be determined not by a single grade formula, but through the full linkage of:

- Fe grade in ore, %,
- concentrate yield, t concentrate/t ore,
- concentrate price, USD/t concentrate,
- transport, USD/t concentrate, and
- mining and beneficiation cost, USD/t ore.

That is the practical meaning of such calculations: not merely to produce a number, but to understand which variables actually determine that number.

3.4.12 How to Read the Key Project Indicators?

For a rapid reading of an iron ore project, it is useful to look not only at Fe grade in ore, but at three things:

- How much concentrate is produced from the ore?
- How much ore must be processed to achieve a given concentrate output?
- And what operating margin remains per tonne of concentrate and per year?

Use the same inputs:

62% Fe concentrate price FOB = **USD 91/t concentrate**
transport of concentrate to port = **USD 10/t concentrate**
mining = **USD 3/t ore**
beneficiation = **USD 15/t ore**
concentrate production = **1.5 Mt/y concentrate**

Net concentrate price:

USD 91/t concentrate − USD 10/t concentrate = **USD 81/t concentrate**

Mining and beneficiation cost:

USD 3/t ore + USD 15/t ore = **USD 18/t ore**

From the previous calculation, take the following concentrate yields:

at 25% Fe: **0.225 t concentrate/t ore**
at 30% Fe: **0.288 t concentrate/t ore**
at 35% Fe: **0.350 t concentrate/t ore**

3.4.13 How Much Ore is Needed for 1.5 Mt/y Concentrate Output?

Formula:

Ore, t ore/y = 1.5 Mt/y concentrate / yield, t concentrate/t ore

Substitution:

at 25% Fe: 1.5 / 0.225 = **6.67 Mt/y ore**
at 30% Fe: 1.5 / 0.288 = **5.21 Mt/y ore**
at 35% Fe: 1.5 / 0.350 = **4.29 Mt/y ore**

3.4.14 What Mining and Beneficiation Cost Per Tonne of Concentrate?

Formula:

Cost per tonne of concentrate, USD/t concentrate = USD 18/t ore / yield, t concentrate/t ore

Substitution:

at 25% Fe: 18 / 0.225 = **USD 80.0/t concentrate**
at 30% Fe: 18 / 0.288 = **USD 62.5/t concentrate**
at 35% Fe: 18 / 0.350 = **USD 51.4/t concentrate**

3.4.15 Operating Margin Per Tonne of Concentrate

Formula:

Operating margin, USD/t concentrate = USD 81/t concentrate − cost per tonne of concentrate

Substitution:

at 25% Fe: 81 − 80.0 = **USD 1.0/t concentrate**
at 30% Fe: 81 − 62.5 = **USD 18.5/t concentrate**
at 35% Fe: 81 − 51.4 = **USD 29.6/t concentrate**

3.4.16 Annual Operating Margin

Formula:

Annual operating margin, USD/y = margin, USD/t concentrate × 1.5 Mt/y concentrate

Substitution:

at 25% Fe: 1.0 × 1.5 = **USD 1.5 million/y**
at 30% Fe: 18.5 × 1.5 = **USD 27.8 million/y**
at 35% Fe: 29.6 × 1.5 = **USD 44.4 million/y**

3.4.17 Summary

TABLE 3.13
Different Heading Grades and Corresponding Results

Indicator	25% Fe	30% Fe	35% Fe
Yield, t concentrate/t ore	0.225	0.288	0.350
Ore required for 1.5 Mt/y concentrate, Mt/y ore	6.67	5.21	4.29
Cost, USD/t concentrate	80.0	62.5	51.4
Operating margin, USD/t concentrate	1.0	18.5	29.6
Annual operating margin, USD million/y	1.5	27.8	44.4

3.4.18 What This Shows Immediately?

At the same concentrate price, the project may be almost empty economically or quite strong.

At 25% Fe, concentrate is still produced, but the project earns almost nothing: only USD 1.5 million per year. At 30% Fe, the economics already work. At 35% Fe, the project looks very different: higher margin and much less ore to process. In other words, one should look not only at concentrate price, but at the full linkage:

Head grade in plant feed → yield → ore tonnage → cost per tonne of concentrate → annual margin

3.4.19 How Time Affects Project Value?

In mining economics, time does not affect project value by itself, but through very concrete factors: mining sequence, speed of ramp-up to nameplate capacity, changes in ore quality by year, strip ratio, and the market phase in which the project receives its main cash flows.

That is why, in practical evaluation, it is important to look not only at the total result over life of mine, but also at when the strongest and weakest cash flows actually occur.

Below are three simple examples.

Assume the Mine Must Extract Two Zones:

- a richer zone generating **USD 70 million/y** of operating cash flow, and
- a poorer zone generating **USD 30 million/y** of operating cash flow.

Assume each zone is mined for two years. The discount rate is **10%/y**, or 0.10.

Option A. Richer zone first
Year 1: USD 70 million
Year 2: USD 70 million
Year 3: USD 30 million
Year 4: USD 30 million

Discounting formula:

PV = CF / (1 + r)^n

Substitution:

Year 1: 70/1.10 = **USD 63.6 million**
Year 2: $70/1.10^2$ = **USD 57.9 million**
Year 3: $30/1.10^3$ = **USD 22.5 million**
Year 4: $30/1.10^4$ = **USD 20.5 million**

Total present value:

63.6 + 57.9 + 22.5 + 20.5 = **USD 164.5 million**

Option B. Poorer zone first
Year 1: USD 30 million
Year 2: USD 30 million
Year 3: USD 70 million
Year 4: USD 70 million

Substitution:

Year 1: $30/1.10$ = **USD 27.3 million**
Year 2: $30/1.10^2$ = **USD 24.8 million**
Year 3: $70/1.10^3$ = **USD 52.6 million**
Year 4: $70/1.10^4$ = **USD 47.8 million**

Total present value:

27.3 + 24.8 + 52.6 + 47.8 = **USD 152.5 million**

3.4.20 What This Example Shows?

The difference between the two options is:

164.5 − 152.5 = **USD 12.0 million**

It is the same mine. The total amount of cash flow is the same. But the project in which the stronger zones generate money earlier is clearly more valuable.

3.4.21 Different Ramp-Up Speeds

Now consider another typical mining example.

Assume the project, once operating, is expected to generate:

Year 1: USD 60 million
Year 2: USD 60 million
Year 3: USD 60 million
Year 4: USD 60 million

Discount rate: **10%/y**
Option A. Normal ramp-up
Substitution:

Year 1: $60/1.10$ = **USD 54.5 million**
Year 2: $60/1.10^2$ = **USD 49.6 million**
Year 3: $60/1.10^3$ = **USD 45.1 million**
Year 4: $60/1.10^4$ = **USD 41.0 million**

Total present value:

54.5 + 49.6 + 45.1 + 41.0 = **USD 190.2 million**

Option B. Slow ramp-up
Now assume the project ramps up more heavily:

Year 1: USD 20 million
Year 2: USD 40 million

Year 3: USD 60 million
Year 4: USD 60 million
Year 5: USD 60 million

Substitution:

Year 1: 20/1.10 = **USD 18.2 million**
Year 2: $40/1.10^2$ = **USD 33.1 million**
Year 3: $60/1.10^3$ = **USD 45.1 million**
Year 4: $60/1.10^4$ = **USD 41.0 million**
Year 5: $60/1.10^5$ = **USD 37.3 million**

Total present value:

18.2 + 33.1 + 45.1 + 41.0 + 37.3 = **USD 174.7 million**

3.4.22 What This Example Shows?

The difference is:

190.2 − 174.7 = **USD 15.5 million**

Even though the total money generated over all years remains broadly comparable, delayed ramp-up reduces project value.

3.4.22.1 Different Market Windows

Now consider an example where time affects the project through the market.

Assume the mine generates **USD 80 million/y** in the first two years and **USD 40 million/y** in the next two years when the metal price is high. At low price, the pattern reverses: the first two years generate only **USD 40 million/y**, and the later years **USD 80 million/y**.

The discount rate is again **10%/y**.

Option A. High price at the beginning of the project

Year 1: USD 80 million
Year 2: USD 80 million
Year 3: USD 40 million
Year 4: USD 40 million

Substitution:

Year 1: 80/1.10 = **USD 72.7 million**
Year 2: $80/1.10^2$ = **USD 66.1 million**
Year 3: $40/1.10^3$ = **USD 30.1 million**
Year 4: $40/1.10^4$ = **USD 27.3 million**

Total present value:

72.7 + 66.1 + 30.1 + 27.3 = **USD 196.2 million**

Option B. High price only in later years

Year 1: USD 40 million
Year 2: USD 40 million
Year 3: USD 80 million
Year 4: USD 80 million

Substitution:

Year 1: $40/1.10 =$ **USD 36.4 million**
Year 2: $40/1.10^2 =$ **USD 33.1 million**
Year 3: $80/1.10^3 =$ **USD 60.1 million**
Year 4: $80/1.10^4 =$ **USD 54.6 million**

Total present value:

$36.4 + 33.1 + 60.1 + 54.6 =$ **USD 184.2 million**

3.4.23 What This Example Shows?

The difference is:

$196.2 - 184.2 =$ **USD 12.0 million**

That is, the same project may have different value not only because of ore quality and mining sequence, but also because of the market phase in which its strongest years occur.

The practical meaning is this: **time affects a project not only through discounting, but also through whether the project's best years coincide, or fail to coincide, with the best years of the market/**

3.4.24 What Should be Seen in Practice?

When reading a mining project, it is useful to ask several simple questions:

- Which years generate the project's main cash flow?
- Which zones are mined early, and which later?
- When does the project reach nameplate capacity?
- How does ore quality change by year?
- Where are the periods of high stripping, difficult processing, or poor quality?
- And in what market phase does the project receive its best tonnes, best grades, and strongest cash flows?

Those questions reveal how time affects project value in real mining logic, not only in an abstract financial formula.

3.4.24.1 Practical Conclusion

In mining economics, time affects project value through:

- mining sequence,
- speed of ramp-up to nameplate capacity,
- distribution of ore quality over time,
- and the coincidence, or lack of coincidence, between the project's strongest years and the market's strongest years.

That is why project evaluation should consider not only the total result over life of mine, but also exactly when the most valuable cash flows arise.

The same mine may have the same reserves, the same total revenue, and the same total metal content, yet very different value simply because the money arrives at different times.

3.4.25 How the Project Changes When Assumptions Change?

In applied mining evaluation, one of the most common errors lies not in the calculations themselves, but in the way the problem is framed. A project rarely behaves like a set of independent parameters. In practice, it is a system of interlinked variables, and that is what must remain in view.

Price affects not only revenue, but also cut-off grade, the economic shell, mining sequence, and mine life. Recovery affects not only the metal content of the product, but also block value, quality thresholds, and the robustness of the process route. Ore quality affects not only grade, but also product yield, processing tonnage, plant load, tailings, water, energy, and transport. Logistics affects not only net realized price, but the economic boundary of the project itself.

That is why a project should be read not as a set of separate numbers, but as a system of equations and dependencies that must first be assembled correctly and only then solved.

3.4.25.1 What a Systems Approach Means?

A systems approach begins with a simple idea: every important variable in the project depends on other variables.

Revenue depends on saleable product volume and net realized price.

Saleable product volume depends on ore tonnage, grade, yield, recovery, feed quality, and the chosen process route.

Net realized price depends on market price, product quality, impurities, delivery basis, transport, and commercial deductions.

Costs depend not only on mining and processing scale, but also on depth, stripping, rock hardness, ore quality, energy, water, reagents, plant operating regime, and logistics.

In the most compressed form, the project can be written as:

Revenue = product volume × net realized price
Product volume = ore volume × product yield

or, if the case is metal,

Payable metal = ore volume × grade × recovery × payable share
Operating result = revenue – mining – processing – logistics – other operating costs

The point of those expressions is not the expressions themselves, but the fact that every variable inside them depends on others. That is where system-level reading of a project begins.

3.4.25.2 Where the Error Usually Starts?

In practice, the error usually begins when one variable is pulled out of the system and treated as if it were independent.

Price is taken as though it were the same for any product quality.

Recovery is taken as though it did not depend on ore type, grade, and processing regime.

Processing cost is taken as though it were the same for any feed.

Logistics is treated as an external fixed block, even though it changes the net price of the product.

Ore quality is described in the geological section, but is not carried through into processing and economics.

The result is that the model may look tidy, while already containing an internal contradiction at its foundation.

3.4.25.3 Why One Number Cannot Simply Be Changed in Isolation

Mechanical change of one variable almost always gives an incomplete picture.

If price falls, not only revenue changes. Cut-off grade, the economic shell, the set of blocks mined, mine life, and mining sequence may all change as well.

If recovery falls, not only metal in product decreases. Product quality, processing requirements, flowsheet stability, and in some cases the viable scale of the project may change too.

If ore quality deteriorates, not only grade declines. Concentrate yield may fall, cost per tonne of product may rise, more ore may need to be processed, and annual margin may weaken.

If logistics cost rises, not only net realized price declines. Marketability itself may change, especially if the project was already close to the economic margin.

That is why the useful applied question is not: what happens if price falls by 10%?

A better question is: through which links will this change pass through the system, and where will the project lose resilience first?

3.4.25.4 How to Read a Project as a System

In practical work, it is often more useful to read a project not by report chapter, but by chains of connection:

- What product is actually sold.
- What determines its volume.

- Which price is the real realized price, not simply a market quotation.
- To which base the costs are referred: per tonne of ore, per tonne of concentrate, per tonne of metal, or per annual volume.
- Which variables are inputs and which are derived.
- Which variable is the controlling variable, the one that changes project outcome faster than the others.

In some projects, the main controlling variable is price. In others, product yield. In others, recovery. In others, logistics. In still others, strip ratio or ore quality by year. Until that variable is identified, the project is being read only superficially.

3.4.25.5 A Useful Working Framework

For practical project reading, it is useful to keep the following linkage in mind:

TABLE 3.14
Main Project Indicators

Variable	What It Changes Directly	What It Changes Next	Where Resilience is Most Often Lost
Product price	Revenue per unit of product	Cut-off grade, economic shell, mining sequence, mine life	The project turns out to be resilient only at too high a price
Mining cost	Cost per tonne of ore	Cut-off grade, operating margin, viable depth and mining shell	A high-strip object loses economics quickly
Processing cost	Cost per tonne of ore or product	Boundaries for low-grade ore, feed-quality requirements, plant robustness	The project works only with unrealistically cheap processing
Logistics	Net realized price	Net revenue, product competitiveness, market accessibility	The price remains on paper, but the net price in the project is already weak
Recovery	Valuable component in product	Revenue, product yield, sometimes cost and throughput	The model uses one recovery, but it does not work on real ore
Ore quality	Product yield, recovery, concentrate quality	Revenue, cost per tonne of product, processing volume, plant load	Averaged ore in the model hides a sharp deterioration in real economics
Concentrate or product yield	Product quantity from 1 tonne of ore	Revenue per tonne of ore, cost per tonne of product, annual margin	As ore quality worsens, the project becomes heavier as a production system
Mining sequence	Timing of cash flow	NPV, capital payback, resilience of the early years	Best zones come too late or are exhausted too quickly
Ramp-up speed	Early cash flow	NPV, capital need, project buffer	A formally strong project loses value because of a difficult start-up

It is usually in the transitions between these rows that the real weakness of a project is hidden.

3.4.25.6 What Systems Thinking Shows in Practice?

If one looks only at product price, a project may appear resilient. But if the same situation is read as a system, the picture changes quickly.

Product price may remain unchanged, but if ore quality worsens, yield declines, revenue per tonne of ore falls, cost per tonne of product rises, more ore must be processed to maintain the same output, and the load on the plant, tailings, water, energy, and transport systems increases. On the surface, one variable has changed. In reality, the whole structure has shifted.

The same applies to recovery. A fall in recovery is rarely an isolated event. It usually reflects ore type, changing feed quality, process limitations, or an overly optimistic original assumption. That is why a fall in recovery is almost never just a reduction in revenue. It is a signal that the processing block of the system is performing more weakly than assumed.

3.4.25.7 What a Robust Model Must Contain?

A robust model does not need to be large. But it must be internally coherent.

It must make clear:

- What the saleable product is.
- How its volume is calculated.
- How net realized price is formed.
- How costs are calculated on a comparable basis.
- How operating result is derived.
- And which variables in the system are most sensitive.

If those things are missing, even a beautiful looking spreadsheet creates only an illusion of precision.

3.4.25.8 Practical Conclusion

Applied project evaluation becomes genuinely useful only when individual indicators stop being treated in isolation. Price, cost, recovery, ore quality, product yield, logistics, and time do not exist separately. They function as an interconnected system, and any change in one of them will eventually pass through the entire project.

That is why reliable evaluation begins not with the search for the "right number," but with the correct framing of the problem itself. It is important not only to calculate the result, but to understand which variables create it, which are inputs, which are derived, and where the project loses resilience first.

If that logic is held, even a relatively simple model can already reveal a great deal. If it is lost, then even a complex spreadsheet with dozens of tabs may create nothing more than an illusion of precision.

That is the practical meaning of a systems approach: it helps one see in the project not a set of indicators, but an interconnected structure in which price, geology, technology, market, and operating scheme must function consistently together. It is exactly this structure that it is useful to return to every time a project looks too simple, too attractive, or too resilient on paper.

3.5 CASE STUDIES IN PROJECT EVALUATION

This section presents applied case studies in the evaluation of mining projects. They are structured so that the reader can work through the main stages of project assessment in a logical sequence: from geology and product to revenue, costs, risk, and overall resilience. This order not only makes it easier to see the essence of an asset more quickly, but also helps identify at which stage the project is confirmed and at which stage it should be questioned or screened out.

3.5.1 An Iron Ore Resource That Fails on Phosphorus

Case type: Early rejection of a project at the stage of initial economic screening of the resource.

Resource mineralogy: Complex oolitic structure. Phosphorus at grades of 0.6%–1.1% is present mainly as apatite, very finely disseminated within the concentric layers of hematite and magnetite oolites.

Essence of the case: On iron grade, the deposit looks strong for this ore type: 52%–57% Fe on average. But phosphorus is critically high: typically 0.6%–0.8%, locally up to 1.1%. Already at an early stage, it is clear that the project does not come up against tonnage or Fe but against the quality of the future product. A resource exists; a normal iron ore product from it may not.

Input data:

- Fe in ore: 52%–57%.
- P in ore: 0.6%–0.8%, locally up to 1.1%.
- Ore: oolitic, complex structure.
- Phosphorus mainly as apatite.
- Apatite very finely disseminated in the concentric layers of hematite and magnetite oolites.
- For the purposes of this example, the acceptable P content in a saleable iron ore concentrate is taken as no more than 0.08%, and below 0.05% for a higher-quality product.

Working assumptions:

- Only two standard routes are considered: DSO ore or iron ore concentrate.
- Exotic or unproven deep-dephosphorization schemes are not accepted as a base case at this early stage.
- The project must pass a simple test: can a saleable product be produced that actually sells and leaves a positive economic result?
- If a separate technological "operation" is required for that, the project can no longer be treated as a conventional iron ore project.

3.5.1.1 Core Check Logic

This example is useful because the asset can be screened out at an early stage.

For an object of this type, the first view is not through tonnage and not through high iron grade, but through a sequence of basic checks.

The first question is the hardest and the most important: **can a normal saleable iron ore product be produced on site?** Not the resource as such, not "strong ore on Fe," but the product: one that meets market requirements and can be sold as DSO or as concentrate.

It is at this step that the asset meets its main problem: despite iron grade being high for this ore type, phosphorus here is critically high.

And the phosphorus is not a coarse, relatively easily separable impurity. It sits as finely disseminated apatite in a complex oolitic structure. That means the question is not simply ore quality, but whether a saleable product can be made through standard routes at all:

- Under DSO, material with this phosphorus level does not look like a normal market product.
- Under concentrate, the next question follows immediately: can a standard beneficiation scheme reduce phosphorus to an acceptable level without destroying the technical and economic logic of the project? Already at the early stage, a positive answer is not obvious.

That is why the first conclusion comes before any talk of transport, railway, or port: **on this asset, a normal saleable product is not proven.**

Only once the commercial viability of the product is confirmed does it make sense to move to the second check: logistics. Once it is clear that a product can be produced and sold, then it is time to count the path to port. What will transport cost? Will a dedicated railway be needed? Can the project carry that haulage distance, and, most importantly, does it retain economic sense after all logistics costs?

In this case, that stage of review should not be the main one. Followed correctly, the evaluation sequence is:

resource → ability to produce a saleable product → negative conclusion → stop

And not:

resource → railway → port → logistics → then the question of product quality

The second sequence substitutes the logic. Discussing logistics in the absence of a proven saleable product does not deepen the analysis. On the contrary, it shifts attention away from the main problem. If the product is not proven, then railway, port and export logistics calculations are no longer a continuation of evaluation, but its imitation.

That is where the practical value of this example lies. It shows that an asset can be screened out very early if it is examined in the right sequence: first, the **product**, only then **logistics**. Not the other way round.

3.5.1.2 Conclusion

If an asset does not yield a normal saleable iron ore product on site, then further analysis of logistics, infrastructure and the export route is no longer real project evaluation. It is only an attempt to discuss the path to market for a product that is not yet there.

3.5.1.3 Attempting to Rescue the Project through Concentrate

After a negative DSO conclusion, the next, apparently natural question usually arises: if the ore cannot be sold in its raw form, can the project be rescued through beneficiation and the production of iron ore concentrate?

It is exactly here that the evaluation has to become stricter, not more optimistic.

If DSO does not work, there is a temptation to move to a "concentrate scenario". In that case, the asset must pass the next check: **how realistically can this material be converted into a saleable concentrate?**

It is not enough to say that iron grade in the ore is high and that "beneficiation is probably possible". That approach is too coarse. At this stage, the evaluator has to put three sequential questions:

- Do technological solutions exist at all that can reduce phosphorus in this type of ore to an acceptable level?
- If such solutions exist in principle, what concentrate can actually be produced on output, not in an ideal flowsheet, but in an industrially credible version?
- Will the market accept such a product, and will it be saleable not only technically, but also commercially?

An asset of this type has to be looked at in exactly this order.

The problem here is not simply high phosphorus as such. What matters far more is how this phosphorus sits in the ore matrix. In this case it is bound up with finely disseminated apatite in a complex oolitic structure. That means the task is not the usual removal of a coarse deleterious impurity. It is a considerably more difficult problem: phosphorus is tightly linked to the iron-bearing fraction of the material, so reducing it to a market-acceptable level may require either very fine liberation, or a complex and unstable processing scheme, or a combination of several stages at once.

So the key practical question is:

- **not "can something be improved,"**

but:

- **"what residual phosphorus levels are realistically achievable in the concentrate?"**

This is the decisive point. The market does not buy a process, and does not buy the effort spent on beneficiation. It buys a product against a defined specification. If, even after a complex flowsheet, the concentrate still contains phosphorus at a level that makes the product undesirable for the main consumers, then the idea of "rescuing the project through concentrate" loses its meaning.

That is why, in this situation, a general statement that "beneficiation is possible" is not enough. A stricter approach is required. The question has to be put in applied form: **what concentrate can actually be produced?**

Not in laboratory idealization. Not from the best sample. Not in a single successful pilot run. But as a product that is:

- reproducible,
- industrially achievable,
- acceptable on recovery,
- acceptable on cost,
- and compliant with the market on residual phosphorus.

If that question has no convincing answer, the concentrate route cannot be regarded as a solution. It remains only a hypothesis about a possible technological rescue of the asset.

This is where the evaluator has to be especially careful. In the practice of mining projects, **it is very easy to substitute the absence of a saleable product with the hope of a future technology**. On paper, it looks like progress: DSO did not work, but now there is "a concentrate option". But if it is not proven what product can actually be produced and whether the market will accept it, that is not a new project route. It is only the transfer of the problem to the next stage of evaluation.

So the question has to be framed not as: **"can beneficiation be attempted?"**

but as: **"does beneficiation turn this resource into a saleable iron ore product?"**

If the answer is unclear, negative, or rests only on unproven assumptions, the project does not become stronger. It only moves out of the category of a conventional iron ore project **into the category of a high-risk technological case**.

That is why, after the DSO rejection, the next conclusion has to be stated plainly: before discussing concentrate as a path to rescue, three things must be **proven** simultaneously:

- That the technology really allows a material reduction of phosphorus.
- That the resulting product has an acceptable residual P content.
- That such a product is genuinely wanted by the market and can be sold as a normal commodity, and not as an exception or a problem material.

If even one of these links is not confirmed, then the concentrate scenario does not rescue the project. It only makes it **more complex, more expensive, and more risky**.

3.5.1.4 Main Conclusion

If DSO does not work, switching to concentrate is not a solution in itself. **First, it must be proven that beneficiation actually turns the resource into a saleable product.** Until that is proven, talk of concentrate is not project logic but technological hope.

3.5.1.5 Final Case Lesson

This case shows why an iron ore resource cannot be assessed by Fe grade alone. The decisive question is whether the material can be converted into a saleable product that meets market requirements at acceptable recovery, cost, and technical risk. In this example**, phosphorus is** not a secondary penalty item that can be dealt with later. It is the **controlling constraint of the project**. Unless testwork demonstrates that phosphorus can be reduced to an acceptable residual level in a reproducible, industrially credible and commercially acceptable concentrate, the resource should not be treated as a conventional iron ore development opportunity. It remains a high-risk technological case, not a normal marketable iron ore project.

3.5.1.6 Practical Decision

At this stage, the correct decision is not to continue with logistics, mine scheduling, or infrastructure optimization. **The correct decision is to stop and require proof of product**. Until a saleable product has been demonstrated, the project has no reliable economic basis.

3.5.2 The Averaging Trap in Polymetallic Ores

3.5.2.1 The Apparent Simplicity

At the level of a resource statement, a polymetallic deposit can look deceptively simple. The table may show tonnes, average grades of copper, gold and silver, contained metal, and a positive economic narrative. On that basis, the project may appear attractive: there is enough mineralized material, the average grades are commercially interesting, and the metal basket seems to support more than one revenue stream.

But this apparent simplicity exists only on paper. Once the material enters the mining and processing chain, the deposit is no longer one average orebody. It becomes a system of different ore types, subtypes and metallurgical behaviours.

The central question is therefore not simply how much copper, gold or silver is contained in the resource. The more important question is how each metal-bearing material type can be mined, classified through grade control, processed and converted into a saleable product.

3.5.2.2 The Real Operating Object

In a complex polymetallic orebody, the mine does not deal with one homogeneous feed. It deals with several ore types that may behave very differently in the plant.

Some material may be sulphide-rich and suitable for flotation. Some may be oxidized or transitional and require a different process route. Some material may carry gold that can be recovered through CIP.(Coal in Pulp) Some may be refractory and require separate treatment, while other material may be non-refractory and relatively straightforward. Some zones may contain silver credits that improve the economics of a concentrate. Other zones may carry impurities, clay, carbonate, pyrite, arsenic, organic carbon, or other components that reduce recovery, increase reagent consumption, affect product quality, or create penalties.

In such a project, grade alone is not enough. The same copper grade can have very different value depending on whether the copper occurs in chalcocite, bornite, chalcopyrite, oxide minerals, or mixed forms. The same gold grade can have very different value depending on whether the gold is free-milling, associated with sulphides, locked in refractory minerals, affected by preg-robbing material, or recoverable only under specific process conditions.

For this reason, the real operating object is not the average resource. The real operating object is the raw-material system.

3.5.2.3 The Raw-Material Map – Geo-Metallurgical Model

A complex orebody needs more than a resource model. It needs a raw-material map that links geology to mining, stockpiling, processing, product quality, and economics.

Such a map should define:

- the main ore types and transitional zones;
- the relevant subtypes within each ore type;
- copper, gold and silver distribution by mineral form;
- oxidation state, sulphide content, and refractory behaviour;
- expected flotation response;
- expected leach or CIP response;
- expected recoveries by ore type and subtype;
- acid consumption, cyanide consumption, or other reagent drivers where relevant;
- deleterious elements and product-quality constraints;
- stockpiling rules;
- blending limits;
- routing rules to each plant section;
- and expected product, recovery, cost, and revenue contribution for each material type.

This is not an administrative complication. It is the operating logic of the project. Without it, the mine may produce tonnes, but it does not produce controlled plant feed.

3.5.2.4 How a Disciplined Operation Handles Complexity?

In a well-controlled operation, geological complexity is not discovered at the plant gate. It is recognized earlier and translated into operating rules.

The geological model defines ore domains. The block model carries attributes that matter for processing, not only total metal grades. Mine planning schedules the ore types deliberately. Grade control identifies material before it is sent to the plant. Stockpiles preserve separation where separation is required, or blend material where blending is technically justified. The processing plant receives feed according to a defined routing logic. The financial model then reflects the actual production mix over time, not an artificial average.

For example, one material type may be routed to flotation to produce a sulphide concentrate. Another may be routed to CIP for gold recovery. Refractory material may be stored, blended within strict limits, treated separately, or excluded from early production. Transitional material may be accepted only within defined feed ratios. High-impurity material may be mined but stockpiled until the correct processing window exists. **Marginal material may be ore in one scenario and waste in another.**

This is how a complex orebody becomes a controllable mining business. The complexity does not disappear. It is made operational.

3.5.2.5 The Mathematics of a Complex Orebody

For this type of asset, the logic must become quantitative. It is not enough to describe the orebody as polymetallic or complex. The project has to translate that complexity into recoverable metal, product streams, cost, risk and value.

The starting point is the qualitative and quantitative description of mineralisation. The project must know not only the total grade of copper, gold or silver, but also how each metal is distributed between ore types, mineral forms and metallurgical domains. In practical terms, the resource has to be decomposed into material classes.

For each material class, the project should define:

- tonnage;
- grade by payable component;
- mineralogical form of each component;
- expected recovery by process route;
- expected product and product quality;
- processing cost;
- mining, stockpiling, and handling cost;
- reagent consumption;
- penalty elements and product deductions;
- and technical risk and sensitivity.

The economic object is therefore not simply:

tonnes × average grade

It is closer to:

ore type × metal form × process route × recovery × product quality × cost × time × risk

This is the applicated mathematics. Geological metal becomes economic metal only after it passes through this chain.

For each ore type, the project should not only ask:

contained metal = tonnes × grade

but then immediately continue:

recoverable metal = contained metal × metallurgical recovery

and then:

payable product value = recoverable metal × payability × realised price – penalties – logistics deductions

Only after that can the project compare revenue with the full cost of producing that product:

cash contribution = payable product value – mining cost – processing cost – logistics cost – sustaining cost – treatment and refining charges – other deductions

In a simple project, this may be manageable with a small number of assumptions. In a complex polymetallic project, the calculation must be performed by ore type, by product route and by mining period. Otherwise, the model hides the very variability that determines value.

This is where the geometallurgical map becomes the mathematical bridge between geology and economics. It should not be a decorative appendix. It should drive the production schedule, stockpile rules, plant feed plan, recovery assumptions and financial model.

A disciplined evaluation should therefore build the asset from the bottom up:

1. define the mineralization qualitatively and quantitatively;
2. separate the deposit into geometallurgical domains;
3. assign realistic recoveries and costs to each domain;
4. define the correct process route for each material type;
5. calculate expected products and product quality;
6. analyze the full chain for each payable component;
7. test sensitivity to recovery, feed mix, reagent consumption, product penalties, costs, and timing;
8. identify which domains create value, which destroy value, and which are only conditionally economic;
9. and evaluate the technical risks before the financial model treats the material as economic ore.

The critical point is that a polymetallic project cannot be evaluated by a single average recovery or a single average cost. The project may be profitable only if the right ore type reaches the right process route at the right time. A small change in feed mix can change recovery, reagent consumption, product quality and revenue simultaneously. That is why sensitivity analysis must be linked to ore type, not only to metal price.

The strongest test is simple: if one ore type is sent to the wrong route, what happens to value? If the answer is lower recovery, higher cost, poor product quality, excessive reagent consumption, or value locked in tailings, then the project depends on ore-type control. In that case, the geometallurgical map is not optional. It is part of the mine design.

3.5.2.6 What Happens When This Logic is Missing

In weak projects, the resource model reports average grades and the financial model applies average recoveries. The mine plan then assumes that material can be mined and fed to the plant as if it were one technological stream. **The plant is expected to absorb the variability then that is a dangerous assumption.**

If sulphide, oxide, transitional, refractory, and non-refractory materials are not identified and managed as separate operating categories, the plant receives geological variability rather than designed feed. Flotation performance becomes unstable. Leaching or CIP response becomes unpredictable. Reagent consumption rises. Recovery assumptions stop matching reality. Concentrate quality may deteriorate. Cathode production or gold recovery may fall short. Tailings and intermediate products may contain value that the flowsheet was supposed to recover but cannot recover under actual operating conditions.

The result is often a project that appears to operate, but does not validate its original economic case. It may produce some concentrate, some doré, some cathode copper, or some saleable product. But partial production is not the same as project validation. It only proves that part of the flowsheet can work under certain conditions. It does not prove that the full resource can be converted into the product mix assumed in the financial model.

3.5.2.7 The Averaging Trap

The most common mistake is to treat contained metal as economic metal too early.

A polymetallic project may show attractive contained copper, gold and silver. But contained metal has value only when it can be recovered into a product that the market will pay for. In a complex orebody, that conversion depends on mineralogy, process route, ore control, timing, blending, recovery, impurities, and cost.

The more complex the orebody, the less meaningful the average grade becomes.

Average grade may be useful for a first geological impression, but it is a poor substitute for operating logic. Average recovery can be even more misleading. If one ore type recovers well and another recovers poorly, the average may conceal the fact that a significant part of the resource is economically weak, process-sensitive or dependent on a process route that has not yet been proven.

A simplified example shows the trap.

Assume a polymetallic operation mines 1.0 Mt of ore in a given period. The average grade appears attractive:

copper: 1.0% Cu,
gold: 0.8 g/t Au,
and silver: 20 g/t Ag.

For illustration, assume current market-based prices of approximately USD 12,900/t Cu, USD 4,750/oz Au and USD 75.7/oz Ag. These are not intended as a long-term price deck. They are used only to show the sensitivity of the project logic.

In the financial model, the ore is treated as one blended feed. **The assumed recoveries are 85% for copper, 80% for gold, and 70% for silver.**

The model therefore shows the following gross recoverable metal value:

- contained copper: 10,000 t Cu;
- recovered copper: 8,500 t Cu;
- copper value: about USD 109.8 million;
- contained gold: about 25,721 oz Au;
- recovered gold: about 20,576 oz Au;
- gold value: about USD 97.8 million;
- contained silver: about 643,015 oz Ag;
- recovered silver: about 450,110 oz Ag;
- silver value: about USD 34.1 million.

Total gross recoverable metal value is therefore about **USD 241.6 million**. If mining, processing, G&A, logistics, treatment charges, sustaining capital and other period costs are modelled at **USD 175 million**, the operation appears to generate about **USD 66.6 million** of operating margin.

But the average model hides the ore-type problem. In reality, **40% of the feed is refractory or transitional material.** It carries a meaningful share of the gold and silver, consumes more reagent, depresses flotation performance, and sends part of the payable metal into tailings or low-quality intermediate products.

Once the actual feed mix reaches the plant, **the effective recoveries fall to 70% for copper, 45% for gold, and 35% for silver. Processing and handling costs also rise because of reagent consumption**, longer residence time, rehandling, extra laboratory control, lower plant stability, reduced throughput, and additional tailings or water-management burden. Total period costs increase from **USD 175 million to USD 195 million**.

The same mined material now produces a very different result:

- recovered copper: 7,000 t Cu;
- copper value: about USD 90.4 million;
- recovered gold: about 11,574 oz Au;
- gold value: about USD 55.0 million;
- recovered silver: about 225,055 oz Ag;
- silver value: about USD 17.0 million.

Total gross recoverable metal value **falls to about USD 162.5 million. Against USD 195 million of costs**, the operating result becomes negative by about **–USD 32.5 million**.

Nothing dramatic changed in the resource table. The tonnes were still mined. The average grades still looked respectable. The metal was still present in the rock. What changed was the conversion of contained metal into payable product, and the cost of forcing the wrong material through the wrong operating system. That is where the project lost value.

The financial model may then monetize material that the operating system has not demonstrated it can process. In that case, the model is not describing the project. It is describing a simplified version of the project that does not exist in the mine.

3.5.2.8 Case Lesson

The failure in this type of project is not that the orebody is complex. Complex orebodies can be mined and processed successfully. The failure occurs when complexity is acknowledged in geological reports but ignored in operating logic.

A polymetallic or multiple ore types orebody is not evaluated by adding metals together. It is evaluated by proving that each metal-bearing material type can be converted into the correct product through the correct route, at the correct time, with realistic recovery and cost.

In a strong operation, ore variability determines the mining strategy, stockpile design, processing route, recovery assumptions, product mix and financial model. In a weak project, ore variability is averaged away. The mine produces material, the plant receives surprises, recovery becomes unstable, costs rise, product quality becomes inconsistent, and the financial model loses contact with reality.

A complex orebody does not fail because it is complex. It fails when the project is designed as if it were simple.

3.5.3 Race for Critical Minerals, Non-Critical Economics

Critical minerals have become one of the strongest narratives in modern mining. Governments want secure supply chains. Investors look for exposure to energy transition, defence, electronics, magnets, batteries, and advanced manufacturing. Companies want to show that their assets are strategically relevant. As a result, many mineral occurrences that would previously have been treated with caution are now presented as critical raw material opportunities.

This is understandable, but dangerous

A mineral may be critical for an economy, but that does not make every occurrence of that mineral economically critical. A rare earth anomaly is not a rare earth project. A monazite-bearing sand is not automatically a business. A tailings pile with elevated critical metals is not automatically a resource. A few million tonnes of mineralized material may sound large in a geological presentation, but may be far too small to support its own mine, plant, hydrometallurgical circuit, product qualification programme and marketing structure.

This is one of the most common mistakes in the current race for critical minerals. People find several hundred thousand tonnes, or a few million tonnes, of material containing rare earths, tungsten, antimony, lithium, cobalt, titanium, zirconium, or another politically attractive element. They calculate contained metal, multiply it by a headline price, and begin to treat the occurrence as an asset and to dream. But they often skip the first industrial question:

Is there enough material, for long enough, at sufficient quality, to justify a production system?

If the answer is no, the project may still be technically interesting. It may be a satellite feed, a by-product opportunity, a research object, or a future option. But it is not yet a standalone mining business.

This is where the evaluation must begin.

The correct sequence is not:

critical mineral → resource → strategic project → investment story

The correct sequence is:

mineral occurrence → scale → continuity → mineralogy → process route → product → buyer → cost → margin → financeability

The first filter is geological scale. Does the occurrence have enough tonnage to matter? This question should not be answered in isolation. It must be linked to the intended business model. A few hundred thousand tonnes may be irrelevant for a standalone mine and plant, but potentially useful as a high-grade campaign feed into an existing operation. Several million tonnes may still be too small if the process route requires a dedicated hydrometallurgical plant, complex reagent systems, residue management, radiological controls, specialized separation, and long customer qualification.

For critical minerals, the threshold tonnage is not a generic number. It depends on grade, recovery, product value, process complexity, mining method, plant capacity, logistics, capital intensity and buyer structure. The practical question is:

Can this material support a stable operating chain, or is it only enough for a short campaign?

A simple early test is to divide mineable tonnes by a realistic annual feed rate.

If a project has 1.5 million tonnes of potentially treatable material and the minimum sensible plant feed is 300,000 tonnes per year, it has only five years of feed before losses, dilution, ramp-up, and selective mining are considered. If the process plant is complex and capital intensive, that is usually not enough. If the same material can be trucked to an existing plant under a tolling or offtake arrangement, the picture may be different. But then the project is not a standalone mine-to-market business. It is a feed source.

That distinction must be made early

The second filter is continuity. Critical mineral stories often rely on scattered samples, old trenches, small lenses, thin zones, residual enrichments, tailings cells, placer pockets or isolated heavy mineral concentrations. This may be enough to justify further work, but not enough to define a production base. A processing plant does not run on enthusiasm. It runs on tonnes, grades, mineralogy, and feed consistency.

The question is not only how much material exists, but whether it can be mined and supplied predictably. Can the material be scheduled? Can the grade be controlled? Is the mineralisation continuous enough to support selective mining? Are the valuable minerals concentrated in mineable domains, or dispersed through waste and low-grade material? **Can the operation deliver a stable feed specification month after month?**

If this cannot be answered, the project remains a geological possibility, not an operating proposition.

The third filter is mineralogy. This is especially important for rare earths. Total rare earth oxide (TREO) content can mislead. A project may report attractive TREO, but the value depends on which rare earth elements are present, in which minerals, and how they can be recovered. A deposit dominated by low-value lanthanum and cerium is different from one with meaningful neodymium, praseodymium, dysprosium, or terbium. Bastnäsite, monazite, xenotime, ionic clay, apatite, allanite, eudialyte, and complex silicate-hosted rare earths are not the same industrial object.

The metallurgical question is blunt:

Can the valuable elements be liberated, concentrated and recovered into a saleable product at acceptable cost?

If rare earths sit in a glassy slag, a complex silicate, a fine-grained refractory matrix, or a low-grade technogenic residue, their analytical presence may have little economic meaning. **The element may be there. The value may not.**

The fourth filter is process route. Many critical mineral projects fail because the process route is assumed rather than demonstrated. In conventional commodities, early assumptions can sometimes be benchmarked against many operating analogues. In REE and other critical minerals, this is often unsafe. The difference between a concentrate, a mixed carbonate, a mixed oxide, separated oxides, metals, alloys or magnet material is not cosmetic. It determines the plant, the buyers, the price basis, the impurity tolerance, the capital intensity, the permitting burden, and the project risk.

A project should not be treated as economic simply because the element has a high market price. it must show a credible route from mined material to product.

The fifth filter is product definition. This is where many critical mineral stories become weak. The company says "rare earths". The market asks: what product? A mineral concentrate? A monazite concentrate? Mixed rare earth carbonate? Mixed oxide? Separated NdPr oxide? A heavy mineral concentrate with monazite as a by-product? A residue stream requiring further treatment?

Each product form has a different buyer universe. Some products are easy to name but hard to sell. Some require qualification. Some carry radionuclide issues. Some depend on a small number of processors. Some are not paid transparently. Some have no realistic market unless attached to a strategic partner.

The sixth filter is buyer access. This is critical. It is not enough to calculate theoretical payable value. The project must answer:

Who will buy this product, in this form, at this scale, with these impurities, from this jurisdiction, under these delivery conditions?

For a small critical mineral occurrence, the answer may be especially difficult. If the project cannot support a standalone plant, then it needs access to existing processing capacity. That means the buyer or processor becomes part of the project logic. Without that buyer, the material may have no route to market. A small monazite stream, for example, may be valuable only if there is a licensed processor willing to take it, pay for its rare earth content, and manage thorium, uranium and downstream separation. Without that route, the "credit" may be imaginary.

The seventh filter is revenue quality. Early-stage critical mineral models often make three mistakes:

1. They use contained metal instead of recoverable product.
2. They use headline prices instead of realized prices.
3. They assume that all valuable elements in the basket will be paid.

This is especially dangerous in rare earths. TREO is not revenue. Contained NdPr is not revenue. Monazite in a concentrate is not revenue unless the buyer pays for it. A by-product credit is not revenue unless it is contractually or commercially credible.

A better early calculation is:

payable revenue per tonne of ore = recoverable product × payable content × realised price minus penalties, discounts, treatment charges, logistics and selling costs

Only after that should the project be compared with the real cost base:

mining + processing + reagents + energy + water + residue management + environmental controls + logistics + G&A + sustaining capital

If the project only works when every element is paid, recovery is high, impurities are ignored, and the buyer accepts the product without discount, it is not an investment case. It is a spreadsheet story.

The eighth filter is capital intensity. This is where small critical mineral projects often die. A small resource may carry a high unit value on paper, but still be unable to support the capital required to process it. A dedicated plant may need crushing, grinding, flotation, leaching, acid handling, impurity removal, precipitation, drying, packaging, water treatment, residue storage, environmental systems, laboratory control, and permitting. For REE, additional complexity may arise from radionuclides, separation, refining, and downstream qualification.

The question is therefore not only whether the material is valuable. The question is whether the value is large enough, long-lived enough, and reliable enough to carry the capital burden.

A simple early-stage test is:

total recoverable gross margin over life of mine ÷ initial capital

If the entire mineable inventory generates only a modest gross margin before sustaining capital, contingencies, ramp-up losses, financing costs, and tax, the project cannot support serious development capital. In that case, the correct strategy may be to stop, sell the material as a niche feed, search for a processor, combine it with other feed sources, or discard the project.

The ninth filter is operating life. Investors do not fund isolated tonnes. They fund a business. A project with three years of feed, uncertain recovery and no product qualification is not a financeable mine, even if the contained metal value looks attractive. A small resource may support a campaign. It may support toll treatment. It may support a blending strategy. But if it cannot support a stable operating life, it should not be marketed as a strategic standalone asset.

This point should be stated clearly:

A critical mineral occurrence must be large enough not only to be mined, but to justify the industrial system required to turn it into product.

The tenth filter is strategic relevance. This is where the current market often gets confused. Government interest, critical mineral lists, supply chain anxiety and geopolitical narratives may improve the chance of grants, strategic investment, offtake support, or policy attention. But they do not replace project economics. Public support can help a marginal but strategically important project. It cannot transform an uneconomic occurrence into a strong asset unless the technical chain is real.

Strategic importance may reduce financing friction. it does not eliminate recovery risk, product risk, cost risk or market risk.

The eleventh filter is timing. Critical minerals are often sold as a response to today's shortage. But if the project needs ten years of exploration, studies, permitting, metallurgical development, plant construction, and product qualification, it may arrive after the market window has changed. **A project must be evaluated against the likely market at the time of production, not the excitement at the time of promotion.**

The final filter is financeability. By this stage, the question should be simple:

> **Can this specific material be mined, processed, converted into a saleable product, delivered to a real buyer, and turned into resilient cash flow at a scale that justifies the capital required?**

If the answer is yes, the project deserves further work. If the answer is no, the word "critical" does not save it.

A practical screening table for an investment banker or technical adviser would look like this:

1. **What exactly is the material?**
   Primary ore, by-product stream, tailings, slag, placer concentrate, industrial residue, or geochemical occurrence?
2. **How much mineable or treatable material exists?**
   Not geological interest, but realistically mineable or recoverable feed.
3. **How many years of operation does it support?**
   Resource tonnes divided by realistic annual feed rate, after dilution, losses, selectivity and ramp-up.
4. **Is the grade meaningful after recovery?**
   Contained grade is only the starting point. Recoverable grade matters more.
5. **Which minerals host the valuable elements?**
   Favourable mineralogy may create a project. Hostile mineralogy may kill it.
6. **Can the material be upgraded?**
   If no concentrate or intermediate can be produced, the route to market is weak.
7. **What product will actually be sold?**
   Not "REE," not "critical minerals," but a defined product with a specification.
8. **Who buys that product?**
   If the answer is theoretical, the revenue is theoretical.
9. **What will the buyer actually pay for?**
   Payability, penalties, deleterious elements, moisture, radionuclides, impurity limits, and treatment charges.
10. **Can the project carry its own plant?**
    If not, is there a credible third-party processing route?

11. **Does the operating margin survive realistic assumptions?**
    Recovery down, cost up, price discounted, by-product credit reduced.
12. **Does the project justify capital?**
    Not just positive margin, but sufficient life, scale, and resilience to support development.

3.5.4.4 The Case Conclusion Should Be Sharp

The race for critical minerals has created a new version of an old mining mistake. In the past, people confused grade with value. Now they confuse criticality with economics. The error is the same. They stop the evaluation too early.

A critical mineral project is not created by the presence of a critical element. It is created by the ability to organize a stable chain from mineralization to product, from product to buyer, and from buyer to cash flow. If the volume is too small, the mineralogy too difficult, the product undefined, the buyer absent, the recovery speculative, or the capital intensity disproportionate, the project should not be dressed up as strategic supply.

It should be stopped, reframed, or put in the bin.

The mineral may be critical. The economics may not be.

4 Synthesis
From Resource to Financeable Project

4.1 WHY SYNTHESIS MATTERS

After the geological foundations, the economic logic, and the practical instruments, the question facing the reader has changed. It is no longer whether a particular discipline has been correctly handled. It is whether the project, taken as a whole, holds together.

Mining projects rarely fail because one discipline is weak in isolation. Geology may be defensible. Mining methods may be known. Process flowsheets may be based on something that has been done elsewhere. Costs can be benchmarked. Markets can be researched. In any single discipline, a competent team can usually produce a respectable answer. The trouble often appears between disciplines, not inside them.

A resource that looks defensible in a geological report may become too narrow when a mining method is applied to it. Mineable material that looks robust in a mine plan may lose tonnes when dilution, selectivity, and haulage route are considered honestly. A process route that performs in the laboratory may not perform on real ore variability at full scale. A product that looks marketable in a presentation may have no buyer willing to accept its specification at the assumed price. A revenue line that works at headline price may collapse at realized netback. A margin that survives operating cost may be destroyed by sustaining capital, ramp-up losses, working capital, and timing. A project that carries positive economics at desk level may still be rejected by capital because the maturity of evidence is too low for the type of money it is asking for.

In each of these cases, the failure is not inside a discipline. It is at the joint between two of them.

That is why this final chapter does not simply summarize what has already been written. The book has already covered geology, mining, processing, product, market, revenue, cost, sensitivity, timing & sequence, project maturity, and financeability. What it now has to do is show how those parts work as one system of evaluation.

Synthesis is not a recap. It is the act of testing whether the chain holds.

The purpose of this chapter is therefore narrow and practical. It converts the book into a working method for asking one question of any mining project: where, in the chain from mineralization to financeability, could this project break? The reader who can answer that question quickly, on evidence rather than narrative, has acquired the main skill this book is trying to teach.

 DOI: 10.1201/9781042018932-4

4.2 THE MINE-TO-MARKET CHAIN AS A CHAIN OF PROOF

It is useful to think of the Mine-to-Market chain not as a sequence of activities, but as a sequence of proofs. Each stage is not only a step. It is a claim that must survive challenge before the next stage has meaning.

The geological object must be real and sufficiently understood. That means continuity, structure, controls, and data quality, not only assays. The resource must have a defensible technical and economic basis. That means tonnage, grade, and continuity that can withstand stricter assumptions, not only a number on a cover page. The material must be mineable under realistic mining conditions. That means geometry, depth, ground conditions, dilution, selectivity, equipment, and people, not only an unconstrained pit shell. The material must be processable. That means a route demonstrated on representative ore variability at appropriate scale, not only a favourable bench-scale composite.

The processed material must become a defined product. That means specification, impurities, deleterious elements, granulometry, moisture, and consistency, not only a name on a flowsheet. The product must have a market. That means buyers who accept that specification at scale, under commercial terms that exist in reality, not only an internal assumption that someone, somewhere, will take it.

Revenue must be realized, not assumed. That means treatment charges, refining charges, payability, penalties, premiums, discounts, logistics, and selling costs, all deducted from the price the analyst saw on a screen. Margin must survive the full cost chain. That means mining, processing, reagents, energy, water, residue management, environmental controls, G&A, sustaining capital, royalties, and tax, including the items that quietly accumulate after the first revenue.

Cash flow must arrive in time and remain reliable. Early ounces, early concentrates and early tonnes are not the same economic event as late ones. Sequencing decides whether the project repays capital before risk catches up with it. Finally, the risk profile must be acceptable to the capital the project is asking for. Acceptable does not mean risk-free. It means suitable for the source of funds, the stage of the project, and the structure of the deal.

A project that has not produced these proofs is not yet a project in the financeable sense. It may be an exploration target with promise. It may be a study with momentum. It may be a story with sponsors. None of that is the same as a chain of evidence.

The chain is what investors, lenders, and operators examine when the conversation becomes serious. A project becomes financeable when the chain holds end to end. It fails when one link remains unproven, no matter how strong the others appear.

4.3 HOW TO USE THE TOOLS IN THIS BOOK

The preceding sections of this book are not subjects sitting next to one another. They are layers of a single evaluation method, and they are most useful when applied in sequence.

First, the geological logic from Chapter 1 is used to test whether the project is real and defensible. The questions are not about deposit type alone. They are about data quality, sample integrity, data density, interpretation, modelling, assumptions and classification. If the geological foundation is weak, every later number inherits that weakness. There is no mine plan, process route, or financial model that can compensate for a resource that is internally unsupported.

Second, the resource and modifying-factor logic is used to test whether mineralized material can credibly become mineable material. Cut-off grade, ore quality, dilution, selectivity, mining method, depth, ground conditions, infrastructure, and timing all sit here. This is the stage at which apparent tonnes are filtered into real tonnes. The reduction is sometimes severe. That severity is not a problem. It is information.

Third, the processing and product-market logic is used to test whether mined material can become a saleable product, and whether that product can be placed with a real buyer at terms the project can survive. This is where many otherwise plausible projects quietly fail. The ore is mineable. The plant is technically feasible. The process route is recognized. But the product, as actually produced, does not match what the market will pay for at the assumed netback.

Fourth, the revenue, cost, sensitivity, timing, and sequencing tools are used to test whether value survives. A project that looks profitable in the base case but cannot withstand a realistic combination of price, recovery, cost, and schedule pressure is not robust. It is fragile. Sensitivity work is not decoration around the financial model. It is where the model is tested.

Fifth, the maturity and financeability tools are used to test whether capital can trust the project. Maturity is not a label attached to a study. It is the level at which critical assumptions have been confirmed by appropriate evidence. Financeability is not a feeling. It is a fit between the project, its risks and the type of capital it is trying to attract.

The checklists in Chapter 3 are not a substitute for judgement. They are a way of disciplining it. Used well, they force the evaluator to stop at each stage, write down what is proven, what is assumed and what is missing, and then decide whether to move forward, refine the model, or stop. Used badly, they become a tick-box exercise.

The difference is not in the checklist. It is in the seriousness of the person using it.

4.4 THE EVALUATION MATRIX

The matrix shown in the Table 4.1 condenses the chain into one page. Each row is a stage. Each stage carries a central question, the evidence required to answer it, and the typical way in which projects fail at that link.

TABLE 4.1
The Evaluation Matrix

Stage	Central Question	Evidence Required	Typical Failure
Geology	Is the object real and sufficiently understood?	Data density, sample integrity, defensible interpretation, deposit-type analogy, internally consistent model	Story driven by a few strong holes; interpretation outruns the data
Resource	Is tonnage, grade, and continuity defensible?	Classification supported by drill spacing, QA/QC, geostatistical behaviour, reasonable prospects of extraction	Tonnes inflated by aggressive interpolation; grades supported by a small population of high samples

(Continued)

TABLE 4.1 *(Continued)*
The Evaluation Matrix

Stage	Central Question	Evidence Required	Typical Failure
Mining	Can the material be extracted under realistic conditions?	Mining method matched to geometry, ground conditions, equipment, dilution, selectivity, and schedule	Unconstrained pit shell; selectivity assumed rather than designed; underground geometry incompatible with the chosen method
Processing	Can the material be converted into a saleable product?	Representative metallurgical testwork across ore types, scale-up logic, recovery confirmed on variability composites	Bench-scale results on best ore extrapolated to the full feed; ore variability acknowledged in geology but ignored in processing
Product	What exactly will be sold?	Defined specification, impurity profile, granulometry, moisture, consistency and packaging	Project sells "concentrate" or "rare earths" without specifying form, grade, or impurity limits
Market	Who buys it, under what specification and commercial terms?	Identified buyer universe, indicative or signed offtake interest, realistic commercial terms	Theoretical market; assumed buyer; price taken from a screen rather than from a contract structure
Revenue	What is the realized netback?	Payability, treatment and refining charges, penalties, premiums, logistics, selling cost, and FX	Headline price treated as revenue; by-product credits assumed without buyer commitment
Costs and cash flow	Does margin survive the full cost chain?	Mining, processing, reagents, energy, water, residue management, G&A, sustaining capital, royalties, tax, and working capital	Operating cost shown; sustaining capital, ramp-up losses, working capital, and tax quietly omitted
Timing and sequence	Does value arrive early enough and in the right order?	Mining sequence designed to bring high-margin tonnes forward, realistic ramp-up schedule, defensible capital phasing	Average-grade, average-cost, average time presentation hiding a back-loaded value profile
Project maturity	Are the assumptions supported by the right level of study?	Stage-appropriate study, independent review, confirmed critical path items	Study label overstates the depth of work behind it; confidence inflated by presentation
Financeability	Can the project withstand capital due diligence?	Resource confidence, technical maturity, market credibility, cost discipline, permitting path, risk allocation, sponsor strength, downside resilience	Project is economically positive in a desk model but cannot answer serious lender questions

The matrix is not a scoring system. It is a structure for honest reading. Two projects with the same headline NPV may sit at very different points in the chain. One may have cleared most of the rows. The other may have cleared only the first two and assumed the rest.

The discipline imposed by the matrix is to refuse to move down the chain faster than the evidence allows. In real evaluation work, this is harder than it sounds. There is constant pressure to discuss markets before processing is settled, to discuss financeability before the resource is robust, to discuss critical mineral status before the product is defined. The matrix slows this down on purpose.

It is also useful as a comparative instrument. In a portfolio review, projects should not be compared only by NPV. They should be compared by how many rows of the matrix they have actually passed. A project that has passed eight rows credibly is in a different state from one that has passed three rows credibly and assumes the next eight. The financial model may not show that difference. The matrix does.

4.5 FOUR LEVELS OF PROJECT QUALITY

Mining projects are often discussed as if they belonged to one undifferentiated category. In reality, they sit at one of four very different levels, and confusing those levels is one of the most common mistakes in early evaluation.

The first level is **geologically interesting**. Mineralization exists, or appears likely to exist, with characteristics worth investigating. There may be drill intercepts, structural indications, geochemical anomalies or encouraging analogies to known deposit types. At this level, the right activities are exploration and geological work. Commercial claims are premature. Talking about NPV(Net Price Value), IRR (Internal Rate of Return), offtake and financing at this level is not ambition. It is noise.

The second level is **technically possible**. There is a plausible route from mineralized material to extracted ore and from extracted ore to a defined product. Mining method, processing approach and product form are no longer purely theoretical. The project may still be weak economically, but the engineering chain is no longer hypothetical. Many projects stop here permanently. They are real in the sense that they could be built, but they cannot generate enough margin under realistic assumptions to justify being built.

The third level is **economically viable**. The project generates positive economics under realistic, not optimistic, assumptions. Recovery is supported by representative testwork. Costs are built up rather than borrowed from a single analogue. Revenue uses realized, not headlined, terms. Sensitivity does not destroy the case. At this level, the project deserves serious work. But it is still not the same as financeable.

The fourth level is **financeable**. The project has enough technical definition, market credibility, cost discipline, permitting path, execution logic, risk allocation, sponsor strength and downside resilience to be acceptable to the type of capital it is trying to attract. Financeability is relative. A project may be financeable for a strategic partner with offtake interest but not for a senior lender. It may be financeable at one level of equity and not another. Financeability is therefore not a single point. It is a fit between the project and a specific source of money.

The key sentence is short and uncomfortable: economically viable is not the same as financeable. A project can pass economic tests on paper and still fail financeability tests in a serious capital conversation. The reasons are familiar by this stage of the book: the resource is not classified at the level required, the metallurgical evidence does not cover the full ore variability, the product specification has not been validated with buyers, the permitting path is unclear, the sponsor cannot carry the risk, or the country risk cannot be absorbed at the implied cost of capital. None of these flaws appears in the operating margin line. All of them appear in the financing conversation.

The four levels also protect the evaluator against premature promotion. A geologically interesting project should not be presented as technically possible. A technically possible project should not be presented as economically viable. An economically viable project should not be presented as financeable. Each promotion has to be earned, and earning it consumes time, money, and evidence.

Compressing the levels in a presentation does not compress them in reality.

4.6 CRITICAL MINERALS, STRATEGIC DEMAND AND REGULATORY REALITY

The current market context cannot be ignored. Critical minerals, energy transition, defence priorities, industrial policy, and supply-chain security have moved several materials into a different conversation. Lithium, copper, nickel, cobalt, rare earths, graphite, tungsten, antimony, niobium, tin, and several others now attract a level of strategic attention that was unusual in earlier cycles. This is not a passing trend. It changes how some projects are looked at, who looks at them, and what kind of capital becomes available.

Strategic relevance can genuinely improve a project. It can support policy backing, public-private financing, prepayment structures, strategic partnerships, and offtake interest. In some jurisdictions, it can support infrastructure investment, accelerate government attention or attract end users that conventional mining finance would not bring in on its own. These are real advantages, and they should be used where they exist.

But strategic relevance does not cancel project economics.

The mineral may be critical. The economics may not be.

This distinction matters because the current market regularly blurs it. A strategic mineral does not automatically mean an economic project. A small occurrence of a critical element is not, on its own, a mining business. The same filters that apply to any project still apply: scale, continuity, mineralogy, process route, product specification, buyer, payable revenue, cost, capital intensity, and life of mine. If those filters are not passed, government interest, lists, narratives, and supply-chain anxiety do not pass them retrospectively.

Likewise, an economically viable project is not automatically a financeable one, even if its commodity appears on every priority list. Financeability still requires resource confidence, a defensible process route, a defined product, a real buyer, cost discipline, a realistic permitting path, a credible sponsor, and resilience to downside cases. Strategic backing may make the financing path easier. It does not replace the underlying chain of proof.

There is also a boundary to this book that should be stated clearly. The preceding chapters have not attempted to analyze regulatory and legal risks in country-by-country detail. That was deliberate. If the book tried to cover every permitting regime, environmental approval process, mining code, land access rule, local content obligation, tax structure, licence category, and government interface across major mining jurisdictions, it would become a different book. The Mine-to-Market framework would be lost inside jurisdictional description.

That does not mean these issues are secondary. They are often decisive. Any serious project evaluation must examine, at the beginning and not at the end, the type of licence held, the rights granted by that licence, the security and duration of tenure, renewal conditions, work commitments, environmental approval requirements, land access, community obligations, water rights, local legislation, tax and royalty regime, export rules, permitting sequence, and the practical attitude of the authorities. A project that is technically strong but legally or environmentally impossible is not a project delayed by paperwork. It is a project without a route to execution.

The correct place for these issues in the framework is not as an afterthought. They belong partly in project maturity and partly in financeability. A resource may be real. The process route may work. The product may be saleable. The economics may be positive. But if the licence does not allow the planned activity, if the environmental approvals cannot be obtained, if land access is unresolved, if water cannot be secured, or if local rules make the assumed development path impossible, the chain breaks before capital can rely on it.

For that reason, the evaluator must treat regulatory, permitting and legal reality as a hard gate. This book does not replace jurisdiction-specific legal, environmental, or permitting due diligence. It makes clear where such due diligence enters the project logic: before financeability is claimed, and before a technical story is allowed to become an investment case.

A useful internal test, when a project is being promoted on strategic grounds, is to read the case as if the commodity were ordinary. If the project still works on its scale, mineralogy, process route, product, buyer, cost, permitting path, and life, then it is a real project that also happens to be strategic. If it only works because the commodity is strategic, then it is a story that depends on subsidies, narratives or buyers who do not yet exist. Both situations can exist. They are not the same thing, and they should not be financed in the same way.

4.7 PRACTICAL USE CASES

The synthesis presented in this chapter is designed to be applied, not admired. It is most useful in the situations in which mining professionals routinely have to make decisions on incomplete information, under pressure and across disciplines.

In **investor screening**, the matrix and the four levels allow a fast triage of opportunities. A project can be placed on the chain within a single review session. The relevant question is not whether it is attractive, but where it currently sits and what would have to be proven to move it to the next level.

In an **owner's internal project review**, the same structure helps test whether the in-house team is honest about its own assumptions. Most internal reviews are not undermined by the lack of data. They are undermined by reluctance to challenge a story that has already been told to the board.

In **lender preparation,** the chain shows where the gaps will be tested first. Lenders rarely fail a project on its strongest discipline. They fail it on its weakest joint. Knowing which joint that is before diligence starts is the difference between an organized process and a defensive one.

In **offtake discussion**, the structure clarifies what the buyer is actually being asked to commit to: a defined product, with a defined specification, at a defined scale, on defined terms. Offtake conversations that take place before product definition is settled are not commercial conversations. They are exchanges of intent.

In **preparation for PFS (Pre-Feasibility Study), FS (Feasibility Study), or financing discussion**, the framework defines the work programme. The critical path is no longer driven by a calendar but by the rows that still need to be closed. This tends to be unpopular with project schedules and useful for project decisions.

In **project planning**, the matrix determines what the next study has to deliver. A study that does not move the project from one row to the next, or from one level of quality to the next, is rarely worth its cost. The decision before commissioning a PFS or FS should be specific: which assumption is being closed, with what evidence, against what threshold?

In **early-stage red flag review**, the matrix is used in reverse. The reviewer looks for the joint at which the chain has not yet been proven and asks whether the project's claim is consistent with that gap. Most red flags are not technical surprises. They are mismatches between the level of the project and the level at which it is being presented.

In **portfolio comparison**, several projects are placed on the same matrix and compared by the number of rows credibly cleared, not by headline NPV or IRR. This usually changes the ranking. Projects that look comparable on financial metrics are often at very different stages of proof.

In **professional assessment of mining managers and executives**, the chain is a diagnostic of integration ability. Mining is taught vertically: geology, mining, metallurgy, processing, commercial, finance, and legal. Most candidates can speak fluently inside one or two columns. The relevant question for senior roles is whether they can move across the chain without losing precision, recognize where value is created or destroyed, and identify the joint at which a project under their responsibility is most likely to fail. A structured walk through the matrix, applied to a real or hypothetical project, exposes that capability quickly. It also exposes its absence.

In **training and professional education**, the same structure works as a teaching backbone. Geologists, engineers, metallurgists, analysts, and finance professionals can be taught their own discipline inside the wider chain rather than in isolation from it. The four levels of project quality, the evaluation matrix, and the logic of proof at each stage give a shared vocabulary across disciplines that normally do not share one. The aim is not to turn specialists into generalists. It is to ensure that each specialist can locate their work inside the system that decides whether the project survives.

4.8 CLOSING PRINCIPLES

A book on mining project evaluation does not need inspirational takeaways. It needs a small number of principles that hold up under pressure. The following are stated briefly because their content has already been earned in the preceding chapters.

Mineralization is not value. The presence of metal, coal or industrial minerals in the ground is a starting condition, not a result. Most mineralized material on the planet will never become a mining project. Treating mineralization as if it were already value is the most common early-stage mistake.

A resource is not a business. A resource is a geological estimate. A business is a system that converts material into product, product into cash flow, and cash flow into return on capital. The distance between the two is paid for in years, money, and discipline.

The product is where geology meets the market. Geology defines the input. The market defines the acceptable output. The product is the only place where those two languages have to agree. If the product is undefined, the project is undefined, regardless of how well its geology is described.

Realized revenue matters more than headline price. Any project that is comfortable at headline price and uncomfortable at realized netback is not robust. It is exposed. The difference between the two numbers is where much early-stage optimism lives.

Costs decide whether value survives. Margin is not a function of revenue alone. It is a function of revenue minus a complete and honest cost chain. Operating cost without sustaining capital, ramp-up losses, working capital, royalties, and tax is not a cost number. It is a fragment.

Time changes value. The same tonne of product, produced early or late, is not the same economic event. Discount rate, debt service, ramp-up risk, and political risk all act through time. A project that is rich in the long run and thin in the short run has a financing problem that no NPV can hide.

Sequencing is an economic instrument. Mining sequence is not only a planning matter. It decides which tonnes repay capital and which tonnes are produced under accumulated risk. Two mining schedules from the same reserve can produce two very different businesses.

Regulatory reality can stop the chain. A project does not become real merely because it is technically possible and economically attractive. Licence type, tenure security, permitting path, environmental obligations, land access, water rights, local law, and government interface can decide whether the project can be executed at all.

Financeability is earned, not assumed. No project is financeable by virtue of its commodity, its geology, or its sponsor. It becomes financeable when its critical assumptions are confirmed at the level required by the type of capital it is asking for. Until then, it is a candidate for financing, not a financeable project.

Weak links matter more than strong narratives. A chain is not measured by its best link. The Mine-to-Market chain is no exception. A project with strong geology, strong product, and strong market, but a weak permitting path or a weak sponsor balance sheet, is not a strong project. It is a fragile project with attractive chapters.

The work this book has tried to support is not the construction of better stories. It is the construction of better judgement. A mining project is real when the chain holds: when the resource is defensible, the material is mineable, the process route is realistic, the product is saleable, the market is real, the margin survives the full cost base, the cash flow arrives in time, the legal and permitting path exists, and the risk fits the capital. When that is true, the project does not need a narrative. The chain is the narrative.

When it is not true, no presentation closes the gap. The discipline of the evaluator is to see the gap clearly, name it without softening it, and decide whether to close, postpone, or stop it. Most of the value created in mining evaluation, over a career, comes from making that decision well. The rest is detail.

That is the use of this book. Not to make projects look better, but to make them be better, or to recognize them honestly when they are not. A resource becomes a financeable project only when the chain is coherent end to end. Everything else in this book has been written in service of that single sentence.

Bibliography/Selected Further Reading

EARTH MATERIALS, MINERALOGY AND PETROLOGY

Klein, C., and Philpotts, A. R. 2016. *Earth Materials: Introduction to Mineralogy and Petrology*. 2nd ed. Cambridge: Cambridge University Press.

ORE DEPOSITS AND ECONOMIC GEOLOGY

Ridley, J. 2013. *Ore Deposit Geology*. Cambridge: Cambridge University Press.

Robb, L. 2020. *Introduction to Ore-Forming Processes*. 2nd ed. Hoboken, NJ: Wiley-Blackwell.

MINERAL RESOURCES, GEOSTATISTICS, AND ESTIMATION

Akin, H., and Siemes, H. 1988. *Praktische Geostatistik: Eine Einführung für den Bergbau und die Geowissenschaften*. Berlin: Springer.

Rossi, M. E., and Deutsch, C. V. 2014. *Mineral Resource Estimation*. Dordrecht: Springer.

Sinclair, A. J., and Blackwell, G. H. 2002. *Applied Mineral Inventory Estimation*. Cambridge: Cambridge University Press.

REPORTING CODES AND RESOURCE DISCLOSURE

CRIRSCO. 2019. *International Reporting Template for the Public Reporting of Exploration Results, Mineral Resources and Mineral Reserves*.

JORC. 2012. *Australasian Code for Reporting of Exploration Results, Mineral Resources and Ore Reserves*.

MINING ENGINEERING AND MINE PLANNING

Darling, P., ed. 2011. *SME Mining Engineering Handbook*. 3rd ed. Englewood, CO: Society for Mining, Metallurgy & Exploration.

Hustrulid, W., Kuchta, M., and Martin, R. 2013. *Open Pit Mine Planning and Design*. 3rd ed. Boca Raton: CRC Press.

Hustrulid, W. A., and Bullock, R. L., eds. 2001. *Underground Mining Methods: Engineering Fundamentals and International Case Studies*. Littleton, CO: Society for Mining, Metallurgy & Exploration.

MINERAL PROCESSING AND EXTRACTIVE METALLURGY

Wills, B. A., and Finch, J. A. 2015. *Wills' Mineral Processing Technology: An Introduction to the Practical Aspects of Ore Treatment and Mineral Recovery*. 8th ed. Oxford: Butterworth-Heinemann.

Dunne, R. C., Kawatra, S. K., and Young, C. A., eds. 2019. *SME Mineral Processing and Extractive Metallurgy Handbook*. Englewood, CO: Society for Mining, Metallurgy & Exploration.

GEOMETALLURGY AND ORE VARIABILITY

Dominy, S. C., ed. 2018. *Geometallurgy: Bridging the Gap between Geology and Mineral Processing*. London: Geological Society, Special Publications.
Lamberg, P. 2023. *Geometallurgy*. Cham: Springer.

MINE ECONOMICS, CUT-OFF GRADE, AND VALUATION

Lane, K. F. 1988. *The Economic Definition of Ore: Cut-off Grades in Theory and Practice*. London: Mining Journal Books.
Rudenno, V. 2019. *The Mining Valuation Handbook: Mining and Energy Valuation for Investors and Management*. 4th ed. Milton, QLD: Wiley.
Guj, P., and Chandra, A. 2019. *Mining Project Evaluation: A Guide to Investment Appraisal*. Melbourne: Australasian Institute of Mining and Metallurgy.

MINERAL ECONOMICS AND COMMODITY MARKETS

Tilton, J. E., and Guzmán, J. I. 2016. *Mineral Economics and Policy*. London: Routledge.
Crowson, P. 2011. *Mineral Economics*. London: Mining Journal Books.

PROJECT FINANCE AND CAPITAL STRUCTURE

Gatti, S. 2018. *Project Finance in Theory and Practice: Designing, Structuring, and Financing Private and Public Projects*. 3rd ed. London: Academic Press.
Nevitt, P. K., and Fabozzi, F. J. 2000. *Project Financing*. 7th ed. London: Euromoney Books.

ENVIRONMENTAL, SOCIAL, AND LENDER REQUIREMENTS

International Finance Corporation. 2012. *Performance Standards on Environmental and Social Sustainability*. Washington, DC: IFC.
Equator Principles Association. 2020. *The Equator Principles EP4*.

COMMODITY AND MARKET REFERENCE SOURCES

U.S. Geological Survey. (Annual). *Mineral Commodity Summaries*. Reston, VA: U.S. Geological Survey.
World Bank. (Annual/quarterly). *Commodity Markets Outlook*. Washington, DC: World Bank.

Index